Isometric Immersions and Embeddings of Locally Euclidean Metrics

CAMBRIDGE SCIENTIFIC PUBLISHERS

Introduction to the series

Volume 12 opens a qualitatively new stage in the development of *Reviews in Mathematics and Mathematical Physics*.

In recent years, many young excellent mathematicians both in Russia and abroad made an excellent name for themselves in the international mathematical community. *Reviews in Mathematics and Mathematical Physics* will publish the most outstanding and recent results. Not only new research, but also a detailed background of a problem will be presented to give an insight into a particular direction. This would enable a better grasp of the state of the art in a particular field and an easier positioning of the results discussed.

To maintain high scientific standards, all works published in *Reviews in Mathematics and Mathematical Physics* are peer reviewed by two or three internationally recognized mathematicians. Many of the works presented by Russian mathematicians have been discussed at the seminars at the Faculty of Mechanics and Mathematics, Moscow State University and Steklov Mathematical Institute, Russian Academy of Sciences.

Reviews in Mathematics and Mathematical Physics also plan publications of review papers by outstanding mathematicians to ensure a broader view of the development of modern mathematics and mathematical physics.

A.T. Fomenko
Editor

Isometric Immersions and Embeddings of Locally Euclidean Metrics

Contents

CONTENTS

CONTENTS

Reviews in Math. & Math. Phys..
2008, Vol. 13, pp. 1–276
Reprints available directly from the publisher
Photocopying permitted by license only

Isometric Immersions and Embeddings of Locally Euclidean Metrics

I.Kh. Sabitov

Faculty of Mechanics and Mathematics, Lomonosov Moscow State University, Leninskiye Gory, Moscow, Russia, 119991
email: isabitov@mail.ru

ABSTRACT

It is known that the major problems of isometric immersions of positive and negative curvature metrics have already been studied in sufficient detail whereas only isolated results are known for zero curvature metrics (i.e. for locally Euclidean metrics). The aim of this work is to review the results on isometric immersions of locally Euclidean metrics into Euclidean spaces along with the description of the extrinsic geometry of such immersions. The review begins with statement of the evidently imposed problem of "natural" realization of locally Euclidean metrics by Euclidean-space domains of corresponding dimension and with the standard Euclidean metric and then studies their isometric immersions into Euclidean spaces of greater dimension with emphasis on the problems of smoothness.

---

*The work was supported in part by grants No. 05-01-00204, No. 09-01-00179 of the Russian Foundation for Basic Research and grant RNP 2.1.1.3704 of the Russian Ministry of Education.

Introduction

It is well known that for two-dimensional metrics of positive or negative curvature the problems of their isometric immersion into E^3 have already been solved in the main (see [1, 7, 8, 25, 26, 32, 69, 124, 127–129, 141]). At the same time, despite the seeming simplicity of the problem, for metrics of zero curvature there are still no sufficiently complete results of their isometric realizations in E^3 in global setting in the form of sufficiently regular surfaces. Besides, in the case of zero-curvature metrics there is a problem specific only to constant-curvature metrics – the realization of locally Euclidean metrics given in an abstract way as a "natural" metric in domains on the Euclidean plane E^2 (for two other classes of constant-curvature metrics, respectively, the problem of their realization by domains on the sphere or Lobachevsky plane). For this reason, we think it useful to consider the results and unsolved problems already available in this field in a unified approach with the aim to attract the attention of geometers engaged in the metric theory of surfaces. Certainly, all problems considered can also be set in multi-dimensional spaces, but there the results are even more scarce than in dimensions 2 and 3. Part of these results will be presented in Chapters 5 and 6, and a general review of the works on the so-called parabolic submanifolds, including, in particular, ruled surfaces, can be found in [24].

1   Regularity of isometries

1.1   Definition of the locally Euclidean metric

A Riemannian metric ds^2 given on an n-dimensional manifold M is called *locally Euclidean* (l. E.) if each point $p \in M$ has a neighborhood U_p isometric to some open ball B_p in E^n with the standard Euclidean metric $d\sigma^2 = (dx^1)^2 + \cdots + (dx^n)^2$. The first problem to be naturally set here is that of the *regularity (smoothness)* of such an isometry. By definition, the isometry between Riemannian manifolds is *a bijective mapping, which preserves the lengths of the curves.* As the regularity of a mapping is its local property, we can confine ourselves to the consideration of local isometries, which is sufficient for studying the regularity of isometries appearing in the definition of the l. E. metric.

4   I.Kh. SABITOV

1.2   *Theorem on the regularity of isometries*

The results on the smoothness of an isometry transferring a given
l. E. metric into the standard Euclidean form will be based on the
following general theorem on the regularity of isometries between
isometric Riemannian manifolds.

Theorem 1.1. *Let two isometric n-dimensional Riemannian spaces
M and N have, respectively, the metrics*

$$ds^2 = g_{ij}(x)dx^i dx^j \text{ and } d\sigma^2 = h_{ij}(y)dy^i dy^j \qquad (1.1)$$

*of smoothness $C^{k,\alpha}$, $0 \leq k \leq \infty$, $0 \leq \alpha \leq 1$, $k + \alpha > 0$. Then, any
isometry f between them has the smoothness of at least class $C^{k+1,\alpha}$.
If isometric Riemannian spaces are analytical, the isometry between
them is also analytical.*

A rather dramatic history of deducing this theorem can be reconstructed in sufficient detail from references in [35] and [144]. The
main mistake in the works before [35] was the use of special coordinates, whose existence or introduction in effect requires (as noted in
[35]) a greater regularity of the metric than it was assumed by the
conditions of the announced theorem. But in [35] itself an elementary
mistake was made too, when estimating the order of decreasing some
infinitesimal function in the last line of p. 745, which was pointed out
in [135]. Therefore, with account for this mistake, the main result
of [35] can be formulated thus: if isometric metrics have a regularity of class $C^{0,\alpha}$, $0 < \alpha < 1$, the isometry f between them should
be at least of class $C^{1,\alpha/2}$. This result was improved in [135] up to
$f \in C^{1,\alpha}$ but it was done using very difficult theorems, the proof of
which makes in essence the dominant bulk of the subject matter of
that book. For this reason, we will do as follows: first we will present
(without proof) a lemma from [35], which is of independent interest.
Based on that lemma, we will prove (following [35]) a weaker version
of Theorem 1.1 (i.e., that $f \in C^{1,\alpha/2}$, if metrics $ds^2, d\sigma^2$ are of $C^{0,\alpha}$).
Next we will prove that in fact $f \in C^{1,\alpha}$ including the case $\alpha = 1$,
and then the remaining part of the theorem is proved simply, the way
it was done in [35].

    Thus, the following lemma (Theorem 3.1 from [35]), which we will
take for granted, is the most important part of the proof.

Lemma 1.1. *In an n-dimensional ($n \geq 2$) Riemannian space V of regularity $C^{0,\alpha}, 0 < \alpha < 1$, the shortest lines $x = x(s)$ are the curves of smoothness $C^{1,\alpha/2}$ with the following estimates:*

$$|x'(s+t) - x'(s)| \leq 2C_0|t|^{\alpha/2} \qquad (1.2)$$
$$|x(L) - x(0) - Lx'(0)| \leq C_0 L^{1+\alpha/2}, \qquad (1.3)$$

where the constant C_0 depends only on the metric of the Riemannian manifold in a sufficiently small domain, and L is the length of the shortest arc considered[1].

Let us comment on the conditions of the lemma.

1) The regularity of Riemannian space V given in the condition means that we can introduce in it some local coordinates (chart) $(x^1, \ldots, x^n)$, relative to which V as a differentiable manifold has the smoothness of class $C^{1,\alpha}$, while the metric is represented by the form

$$ds^2 = g_{ij}(x)dx^i dx^j \qquad (1.4)$$

with the coefficients $g_{ij}(x) \in C^{0,\alpha}$.

2) According to [35], the very existence of the shortest lines is guaranteed by the following proposition ascribed to Hilbert.

Proposition 1.1. *Let the metric form (1.4) have continuous coefficients defined in an open ball $D_1 : |x| < 1$, in which the uniform estimates are satisfied:*

$$\frac{1}{C^2}|\xi|^2 \leq g_{ij}(x)\xi^i\xi^j \leq C^2|\xi|^2, |x| < 1, \forall \xi = (\xi^1, \ldots, \xi^n). \qquad (1.5)$$

Consider a closed ball $\bar{D}_r$ consisting of points with $|x| \leq r < 1$. Then, there exist numbers $\delta = \delta(r) > 0$ and $r_1 = r_1(r,\delta), 0 < r < r_1 < 1$, such that, if points x_0 and x_1 from $\bar{D}_r$ are at a distance (in Euclidean sense) of no more than $\delta : |x_0 - x_1| \leq \delta$ from each other, they can be connected by a line $x = x(s)$ having the shortest length in the metric (1.4) and not falling outside the boundary of the ball D_{r_1}, i.e., satisfying the conditions $|x(s)| \leq r_1$, $0 \leq s \leq L$, with $x(0) = x_0, x(L) = x_1$, where L is the length of the shortest line. Moreover, it can be asserted that at $\delta(r) \to 0$ the value of $r_1 \to r$.

---

[1] In [35], the right-hand sides have α instead of $\alpha/2$, but the error was found namely in the proof of that.

In reality, the work [83], which the authors of [35] appear to have in mind (see the reference to this work in a similar aspect in [79]), gives only an idea of proving a more particular proposition of the existence of a shortest arc between two points on a C^1-smooth surface $z = f(x,y)$. A more detailed proof of this proposition of Hilbert can be found in [22] also cited in [79]. In turn, §57 of [22] as a matter of fact repeats the proof from [104] (pp. 345–348), in an evident way carried over to the case of abstract multidimensional metrics. As to the propositions on the location of shortest arcs inside some ball, they are obtained with the account of inequalities (1.5) making it possible to estimate the deviation of the shortest arcs in the metric (1.4) from those in a respective Euclidean metric.

Note that there is no problem of the existence of shortest arcs as applied to the case when one of the isometric metrics is sufficiently regular (for example, as in our case, when the local Euclidity of the considered metric is known), since they are transferred to a weakly regular metric from a sufficiently regular metric by means of isometry.

We proceed now to the p r o o f of Theorem 1.1. First of all, in addition to Lemma 1.1 we will need further the following:

Proposition 1.2. *The following inequality holds for the length L of the shortest arc in Proposition 1.1:*

$$L \leq C|x(L) - x(0)|. \tag{1.6}$$

Indeed, the length of the shortest arc between $x(0)$ and $x(L)$ is not greater than the length of the chord between these points in the metric (1.4). Setting $x(t) = x(0) + tx(L)$, $0 \leq t \leq 1$, and $\xi = x(1) - x(0)$, we calculate the length of this chord in the metric (1.4) and, taking into account (1.5), we get

$$L \leq \int_0^1 \sqrt{g_{ij}(x(t))\xi^i\xi^j}\, dt \leq C|\xi| = C|x(L) - x(0)|,$$

which is what was proposed.

Let $f : M \to N$ be an isometry between the metrics (1.1). Owing to the validity of the inequalities of the form (1.5) (incidentally, without loss of generality we shall assume these inequalities to hold for both metrics with equal constants C), for both metrics we get that the mapping $y = f(x)$ in corresponding local coordinates satisfies the Lipschitz condition and, thus, by the known theorem of Rademacher [130], it has almost everywhere the differential determined by the Jacobi matrix $F(x) = f'(x) = (f^i_j)$, where $f^i_j = \dfrac{\partial f^i}{\partial x^j}$. Then, the isometric property of the mapping $f(x)$ means that

$$G(x) = F^T(x)H(f(x))F(x), \qquad (1.7)$$

where the matrices $G = (g_{ij})$, $H = (h_{ij})$, and the matrix F^T is obtained by transposing the matrix F.

Let $x = x_0 \in M$ be a point, in which the mapping $f(x)$ is differentiable. Then $df(x_0) = F(x_0)dx$. For an arbitrary vector ξ we take $\eta = F(x_0)\xi$. By (1.5) and (1.7), we get

$$\frac{1}{C^2}|\eta|^2 \le h_{ij}(f(x))\eta^i\eta^j = g_{ij}\xi^i\xi^j \le C^2|\xi|^2,$$

from which we have the following estimate for the norm of the matrix $F(x_0)$:

$$|F(x_0)\xi| \le C^2|\xi|. \qquad (1.8)$$

Consider an arbitrary point $x_1 \in M$ located sufficiently close to x_0. Then, by Proposition 1.1, a shortest arc exists, which connects x_0 with x_1. Let $x = x(s), 0 \le s \le L$, be the equation for this shortest arc. Its image $y = y(s) = f(x(s))$ would be the shortest arc in space N connecting the points $f(x_0)$ and $f(x_1)$. Since $y'(s_0) = F(x_0)x'(0)$, then, using the analogy of the relation (1.3), we get

$$|f(x_1) - f(x_0) - LF(x_0)x'(0)| \le C_0 L^{1+\alpha/2}.$$

Now we take into account the relations (1.3) and (1.8) and get

$$|f(x_1) - f(x_0) - F(x_0)(x_1 - x_0)| \le ||f(x_1) - f(x_0) - LF(x_0)x'(0)| +$$
$$|F(x_0)(x_1 - x_0 - Lx'(0)| = C_0(1 + C^2)L^{1+\alpha/2} = C_1 L^{1+\alpha/2}.$$

Taking into account (1.6), we obtain the final estimate:

$$|f(x_1) - f(x_0) - F(x_0)(x_1 - x_0)| \le C_1 C^{1+\alpha/2}|x_1 - x_0|^{1+\alpha/2}. \quad (1.9)$$

The inequality (1.9) holds for any point x_1 sufficiently close to x_0. Therefore, it can also be written for any other point x_2:

$$|f(x_2) - f(x_0) - F(x_0)(x_1 - x_0)| \leq C_1 C^{1+\alpha/2}|x_2 - x_0|^{1+\alpha/2}.$$

Together with (1.9), this inequality yields that

$$|f(x_2) - f(x_1) - F(x_0)(x_2 - x_1)| \leq$$
$$C_1 C^{1+\alpha/2}(|x_1 - x_0|^{1+\alpha/2} + |x_2 - x_0|^{1+\alpha/2}). \tag{1.10}$$

On the other hand, if the point x_1 is chosen such that $f'(x_1)$ also exists in it, for the points x_1 and x_2 we can write an inequality similar to (1.9):

$$|f(x_2) - f(x_1) - F(x_1)(x_2 - x_1)| \leq C_1 C^{1+\alpha/2}|x_2 - x_1|^{1+\alpha/2},$$

so that, with regard for it and (1.10), we get

$$|(F(x_0) - F(x_1))(x_2 - x_1)| \leq$$
$$C_1 C^{1+\alpha/2}\left(|x_2 - x_1|^{1+\alpha/2} + |x_1 - x_0|^{1+\alpha/2} + |x_2 - x_0|^{1+\alpha/2}\right).$$

Choosing x_2 such that the triangle $x_0 x_1 x_2$ is equilateral (in Euclidean sense) and denoting the unit vector in the direction $x_1 x_2$ as ξ so that $x_2 - x_1 = t\xi$, we have from the previous inequality:

$$|(F(x_0) - F(x_1))t\xi| \leq 3C_1 C^{1+\alpha/2}t|x_1 - x_0|^{\alpha/2}|\xi|.$$

Reducing by $t = |x_0 - x_1| = |x_0 - x_2|$, we arrive at the conclusion that the matrix $F(x) = f'(x)$ at the points of its existence is uniformly continuous with the Hölder exponent of continuity $\alpha/2$. Then it can be extended to a continuous function in a neighbourhood of x_0 with the same Hölder exponent of continuity $\alpha/2$. This proves that the isometry is of the class of regularity $C^{1,\alpha/2}$.

Before passing to the proof that for isometric Riemannian spaces of class $C^{0,\alpha}$ the regularity of isometry would in fact be of class $C^{1,\alpha}$, we shall prove the following two lemmas.

Lemma 1.2. *Let the function* $w(z) = u + iv$, $z = x + iy$, *determined in a circle* $\omega : |z| \leq R$, *satisfy the following conditions:*

$$1)\ w(z) \in C^{1,\alpha/2}(\omega), 0 < \alpha < 1,$$

$$2)\ \frac{\partial w(0)}{\partial \bar{z}} = \frac{\partial w(0)}{\partial z} = 0,$$

$$3)\ \frac{\partial w}{\partial \bar{z}} = f(z), \tag{1.11}$$

where $f(z) \in C^{0,\alpha/2}(\omega), |f(z)| \leq C|z|^{\alpha}, z \in \omega$. *Then for any circle* $|z| \leq r_0 < R$ *there exists a constant* C_1, *which depends on* R, r_0, C, α *and on* $\max|w(t)|$ *on the circumference* $\partial\omega : |z| = R$, *such that*

$$|w(z) - w(0)| \leq C_1|z|^{1+\alpha}, \ |z| \leq r_0. \tag{1.12}$$

Lemma 1.3. *Let the function* $w(z) = u + iv$, $z = x + iy$, *determined in the circle* $\omega : |z| \leq R$, *satisfy the following conditions:*

$$1)\ w(z) \in C^{1,\alpha}(\omega), 1/2 < \alpha < 1,$$

$$2)\ \frac{\partial w(0)}{\partial \bar{z}} = \frac{\partial w(0)}{\partial z} = 0,$$

$$3)\ \frac{\partial w}{\partial \bar{z}} = f(z), \tag{1.13}$$

$$4)\ Re\frac{\partial w}{\partial z} = g(z), \tag{1.14}$$

where $f(z), g(z) \in C^{0,\alpha}(\omega), |f(z)| \leq C|z|, |g(z)| \leq C|z|, z \in \omega$. *Then for any circle* $|z| \leq r_0 < R$ *there exists a constant* C_2, *which depends on* R, r_0, C, α *and on the norm* $w(z)$ *in class* $C^{1,\alpha/2}(\omega)$, *such that*

$$|w(z) - w(0)| \leq C_2|z|^2, \ |z| \leq r_0. \tag{1.15}$$

Proof of Lemma 1.2. The general solution of the equation (1.11) has the form (see pp. 55–56 in [184]) $w(z) = F(z) + \Phi(z)$, where

$$F(z) = -\frac{1}{\pi} \iint_{\omega} \frac{f(\zeta)}{\zeta - z} d\xi d\eta, \zeta = \xi + i\eta, \tag{1.16}$$

$$\Phi(z) = \frac{1}{2\pi i} \oint_{\partial\omega} \frac{w(t)}{t - z} dt = a_0 + a_1 z + a_2 z^2 + \dots. \tag{1.17}$$

We have $w(0) = -\dfrac{1}{\pi} \iint\limits_{\omega} \dfrac{f(\zeta)}{\zeta} d\xi d\eta + \Phi(0)$. In the case when, as in our case, the function f is of Hölder class, also valid is the formula (see Chapter 1, §8 in [184])

$$\frac{\partial w}{\partial z} = -\frac{1}{\pi} \iint\limits_{\omega} \frac{f(\zeta)}{(\zeta - z)^2} d\xi d\eta + \Phi'(z),$$

whence

$$\frac{\partial w(0)}{\partial z} = -\frac{1}{\pi} \iint\limits_{\omega} \frac{f(\zeta)}{\zeta^2} d\xi d\eta + \Phi'(0) = 0.$$

Therefore,

$$a_1 = \Phi'(0) = \frac{1}{\pi} \iint\limits_{\omega} \frac{f(\zeta)}{\zeta^2} d\xi d\eta.$$

Now, with provision for all these relations, we have

$$w(z) - w(0) = -\frac{z}{\pi} \iint\limits_{\omega} \frac{f(\zeta)}{\zeta(\zeta - z)} d\xi d\eta + a_1 z + a_2 z^2 + \ldots$$

$$= -\frac{z^2}{\pi} \iint\limits_{\omega} \frac{f(\zeta)}{\zeta^2(\zeta - z)} d\xi d\eta + a_2 z^2 + \ldots . \quad (1.18)$$

The estimate

$$|a_2 z^2 + \ldots| = |\Phi(z) - \Phi(0) - \Phi'(0)z| = \left| \frac{z^2}{2\pi i} \oint\limits_{\partial \omega} \frac{w(t)}{t^2(t - z)} dt \right| \leq C_6 |z|^2$$

$$(1.19)$$

in any circle $|z| \leq r_0 < R$ is evident; the constant C_6 depends only on r_0, R and on the maximum of $|w(t)|$ on the circumference $|z| = R$. Let us estimate the integral

$$A = -\frac{1}{\pi} \iint\limits_{\omega} \frac{f(\zeta)}{\zeta^2(\zeta - z)} d\xi d\eta. \quad (1.20)$$

For this, we recall the estimate from Chapter 1, §6 [184]:

$$\frac{1}{\pi} \iint\limits_{D} \frac{d\xi d\eta}{|\zeta - z_1|^\gamma |\zeta - z_2|^\beta} \leq K |z_1 - z_2|^{2-\gamma-\beta}, \ \gamma < 2, \ \beta < 2, \ \gamma+\beta > 2,$$

$$(1.21)$$

the constant K depending only on the region D and the constants β, γ. By the condition of the lemma, $|f(z)| \leq C|z|^\alpha$, so in our case for the integral A in the estimate (1.21) we can take $\gamma = 2 - \alpha, \beta = 1, z_1 = 0, z_2 = z$ and obtain that

$$|A| = \left|\frac{1}{\pi} \iint\limits_\omega \frac{f(\zeta)}{\zeta^2(\zeta - z)} d\xi d\eta\right| \leq CK|z|^{\alpha-1},$$

which, together with (1.18) and (1.19), gives the proposition of the lemma.

Proof of Lemma 1.3. We write again the representation of the solution of equation (1.13) by the above used formula from [184]. As the function $f(z)$ belongs to $C^{0,\alpha}, \alpha > 1/2$, on each circumference $|z| = r$ it expands into a Fourier series uniformly convergent on $|z| = r$:

$$f(z) = \sum_{n=-\infty}^{+\infty} f_n(r)e^{in\varphi}, z = re^{in\varphi}, \tag{1.22}$$

where

$$f_n(r) = \frac{1}{2\pi} \oint\limits_{|z|=r} f(z)e^{-in\varphi} d\varphi. \tag{1.23}$$

The remainder of the series permits the estimate (see p. 208 in [12])

$$|f(z) - s_m(z)| \leq C_3/m^{\alpha-1/2},$$

where the constant C_3 depends on r, the exponent α and the Hölder coefficient of the function $f(z)$ on $|z| = r$ with $C_3(r) \to 0$ at $r \to 0$; therefore, the series (1.22) converges to $f(z)$ uniformly in all the circle ω, so with provision for the estimates from [184] for the integral $F(z)$ in (1.16) we get that the series (1.22) under the sign of the integral $F(z)$ permits the termwise integration. Because of this,

$$w(z) = -2\sum_{n=1}^{\infty} z^{n-1} \int_r^R \frac{f_n(\rho)}{\rho^{n-1}} d\rho + 2\sum_{n=0}^{\infty} \frac{1}{z^{n+1}} \int_0^r f_{-n}(\rho)\rho^{n+1} d\rho + \Phi(z)$$

$$\tag{1.24}$$

12   I.Kh. SABITOV

(to obtain this expression, the integral over the domain ω should be split into two integrals – over the circle $|\zeta| \le r = |z|$ and over the ring $r \le |\zeta| \le R$).

Further, the Fourier series for $w'_\varphi = iz\dfrac{\partial w}{\partial z} - i\bar{z}\dfrac{\partial w}{\partial \bar{z}}$ over the circumferences $|z| = r$ is obtained by termwise differentiation of the series (1.24); it also converges uniformly to w'_φ. With provision for the equality $\dfrac{\partial}{\partial z} = \dfrac{\bar{z}}{z}\dfrac{\partial}{\partial \bar{z}} - \dfrac{i}{z}\dfrac{\partial}{\partial \varphi}$, we get

$$\frac{\partial w}{\partial z} = -2\sum_{n=2}^{\infty}(n-1)z^{n-2}\int_{r}^{R}\frac{f_n(\rho)}{\rho^{n-1}}d\rho + \sum_{n=1}^{\infty}\frac{z^{n-2}}{r^{n-2}}f_n(r) -$$

$$-2\sum_{n=0}^{\infty}\frac{n+1}{z^{n+2}}\int_{0}^{r}f_{-n}(\rho)\rho^{n+1}d\rho + \sum_{n=0}^{\infty}\frac{r^{n+2}}{z^{n+2}}f_{-n}(r) + \Phi'(z). \quad (1.25)$$

From (1.25), we have

$$\frac{1}{2\pi}\oint_{|z|=r}\frac{\partial w}{\partial z}e^{-i\varphi}d\varphi = -4r\int_{r}^{R}\frac{f_3(\rho)}{\rho^2}d\rho + f_3(r) + ra_2,$$

$$\frac{1}{2\pi}\oint_{|z|=r}\overline{\frac{\partial w}{\partial z}}e^{-i\varphi}d\varphi = \overline{f_1(r)}.$$

We add up these equalities; then, with provision for the conditions of the lemma, we have

$$\left|\int_{r}^{R}\frac{f_3(\rho)}{\rho^2}d\rho\right| \le C_4.$$

where the constant C_4 does not depend on r. By calculating the integral

$$J_r = -\frac{1}{\pi}\iint_{\omega_r}\frac{f(\zeta)}{\zeta^3}d\xi d\eta, \quad \omega_r : r \le |\zeta| \le R,$$

with $f(\zeta)$ represented in the form of the series (1.22), we get

$$J_r = 2\pi \int\limits_r^R \frac{f_3(\rho)}{\rho^2}\,d\rho.$$

Therefore,

$$|J_r| = \left| \frac{1}{\pi} \iint\limits_{\omega_r} \frac{f(\zeta)}{\zeta^3}\,d\xi d\eta \right| \le C_5 \,(C_5 = 2\pi C_4) \tag{1.26}$$

independently on $r, 0 < r < R$. This is the main nontrivial estimate used in further consideration.

Let us establish the proposed estimate (1.15). For convenience of reading, let us repeat the already known formula

$$w(z) - w(0) = -\frac{z}{\pi} \iint\limits_{\omega} \frac{f(\zeta)}{\zeta(\zeta - z)}\,d\xi d\eta + a_1 z + a_2 z^2 + \dots$$

$$= -\frac{z^2}{\pi} \iint\limits_{\omega} \frac{f(\zeta)}{\zeta^2(\zeta - z)}\,d\xi d\eta + a_2 z^2 + \dots \tag{1.27}$$

and the estimate

$$|a_2 z^2 + \dots| =\le C_6 |z|^2. \tag{1.28}$$

We show that this time the integral A from (1.20) is uniformly bounded for all $z \neq 0$. This integral is understood to be the limit

$$A = \lim_{\varepsilon \to 0} A_\varepsilon = \lim_{\varepsilon \to 0} \left(-\frac{1}{\pi} \iint\limits_{\omega_\varepsilon} \frac{f(\zeta)}{\zeta^2(\zeta - z)}\,d\xi d\eta \right), \quad \omega_\varepsilon : \varepsilon \le |\zeta| \le R.$$

For A_ε, we have the following representation:

$$A_\varepsilon = J_\varepsilon - \frac{z}{\pi} \iint\limits_{\omega_\varepsilon} \frac{f(\zeta)}{\zeta^3(\zeta - z)}\,d\xi d\eta \equiv J_\varepsilon + B_\varepsilon.$$

The term J_ε is bounded on the strength of (1.26). Let us show that the term B_ε is also uniformly bounded for all $\varepsilon > 0$ and for all $z \in \omega, z \neq 0$.

Consider all possible cases of the location of the point z in the circle ω.

1) $|z| \leq \varepsilon$. Then in ω_ε we have $|z| \leq |\zeta|$, so with provision for the inequality (1.21) we get

$$|B_\varepsilon| \leq \frac{|z|^{1/2}}{\pi} \iint\limits_{\omega_\varepsilon} \frac{C\,d\xi d\eta}{|\zeta|^{3/2}|\zeta - z|} < \frac{C|z|^{1/2}}{\pi} \iint\limits_{\omega} \frac{d\xi d\eta}{|\zeta|^{3/2}|\zeta - z|} \leq C_7(C, R).$$

2) $\varepsilon < |z| \leq R$. In this case, we represent the integral B_ε over ω_ε as two integrals: over the ring $K_{\varepsilon,r} : \varepsilon \leq |\zeta| \leq r = |z|$ and over the ring $\omega_r : r = |z| \leq |\zeta| \leq R$. The integral over ω_r permits the just now established estimate via $C_7(C, R)$, and the integral over $K_{\varepsilon,r}$ is transformed as follows:

$$-\frac{z}{\pi} \iint\limits_{K_{\varepsilon,r}} \frac{f(\zeta)}{\zeta^3(\zeta - z)}\,d\xi d\eta = \tag{1.29}$$

$$-\frac{1}{\pi} \iint\limits_{K_{\varepsilon,r}} \frac{f(\zeta)}{\zeta^2(\zeta - z)}\,d\xi d\eta + \frac{1}{\pi} \iint\limits_{K_{\varepsilon,r}} \frac{f(\zeta)\,d\xi d\eta}{\zeta^3}.$$

We have

$$\left| \frac{1}{\pi} \iint\limits_{K_{\varepsilon,r}} \frac{f(\zeta)}{\zeta^3}\,d\xi d\eta \right| \leq \left| \frac{1}{\pi} \iint\limits_{\omega_\varepsilon} \frac{f(\zeta)\,d\xi d\eta}{\zeta^3} \right| + \left| \frac{1}{\pi} \iint\limits_{\omega_r} \frac{f(\zeta)\,d\xi d\eta}{\zeta^3} \right| \leq 2C_5.$$

The first integral in the right-hand side of (1.29) shall be represented as

$$-\frac{1}{\pi} \iint\limits_{K_{\varepsilon,r}} \frac{f(\zeta)\,d\xi d\eta}{\zeta^2(\zeta - z)} =$$

$$\frac{1}{z}\left(-\frac{1}{\pi} \iint\limits_{K_{\varepsilon,r}} \frac{f(\zeta)\,d\xi d\eta}{\zeta(\zeta - z)} + \frac{1}{\pi} \iint\limits_{K_{\varepsilon,r}} \frac{f(\zeta)}{\zeta^2}\,d\xi d\eta \right),$$

whence

$$\frac{1}{\pi|z|}\iint\limits_{K_{\varepsilon,r}}\frac{|f(\zeta)|d\xi d\eta}{|\zeta||\zeta-z|}\leq\frac{C}{\pi|z|}\iint\limits_{K_{\varepsilon,r}}\frac{d\xi d\eta}{|\zeta-z|}<$$

$$\frac{C}{\pi|z|}\iint\limits_{|\zeta|\leq r}\frac{d\xi d\eta}{|\zeta-z|}=4C,$$

$$\frac{1}{\pi|z|}\left|\iint\limits_{K_{\varepsilon,r}}\frac{|f(\zeta)|}{\zeta^2}d\xi d\eta\right|\leq\frac{2C}{|z|}(r-\varepsilon)<2C.$$

As a result, we have that the integral A_ε is bounded uniformly relative to $\varepsilon>0$ and $z\neq0$, so the limit A_ε at $\varepsilon\to0$ and fixed $z\neq0$ is bounded by the constant not depending on z, which, together with (1.27) and (1.28), yields the required. The lemma is proved.

Now we can proceed to continue with the proof of Theorem 1.1. As it is sufficient to establish the statement of smoothness "in the small", we shall consider M and N to be balls $U:|x|<R$ and $V:|y|<R$. First we shall show that with isometric metrics of smoothness of class $C^{0,\alpha},0<\alpha<1$, for any two points a and b of some closed ball $\omega:|x|\leq r_0<R$, the inequality

$$\Delta=|f(b)-f(a)-f'(a)(b-a)|_{h(f(a))}\leq C_0\left(|b-a|_{g(a)}\right)^{1+\alpha}\quad(1.30)$$

holds with some constant C_0 not depending on the choice of points a and b in ω, where on the left-hand side the length of the vector is taken in the Euclidean metric determined by the metric $d\sigma^2$ in point $f(a)\in V$, and on the right-hand side the length is taken in the Euclidean metric ds^2 in point $a\in U$ (the class $C^{1,\alpha/2}$ of isometry $f:U\to V$ is already known from the first part of the proof).

Generally speaking, to obtain the estimate of (1.30), we shall use linear basis transformations both in U and in V with the transfer of the origin, as well as perform some orthogonal transformations over the components of the isometric mapping f. But it is easy to verify that transformations like these change neither the left-hand side nor right-hand side of formula (1.30). For this reason, we can consider that the following conditions are satisfied in the estimation of Δ:

1) $a=0$, $f(a)=0$, and point b has the coordinates $(b^1,0,\ldots,0)$, i.e., is located on the axis Ox^1;

2) $g_{ij}(x) = \delta_{ij} + G_{ij}(x), |G_{ij}(x)| \leq C|x|^\alpha$;

3) $h_{ij}(y) = \delta_{ij} + H_{ij}(y), \; |H_{ij}(f(x))| \leq C|x|^\alpha$;

4) $f^i(x) = x^i + F^i(x)$, so that $\dfrac{\partial f^i}{\partial x^j} = \delta_i^j + \dfrac{\partial F^i}{\partial x^j}, \dfrac{\partial F^i(0)}{\partial x^j} = 0$ and $\dfrac{\partial F^i}{\partial x^j} \in C^{0,\alpha/2}$.

Conditions 2) and 3) are satisfied, if required, by linear transformations of the coordinates, which depend only on the values of $g_{ij}(a)$ and $h_{ij}(f(a))$; for condition 1) to be met, we shall need to transfer the origins of the coordinates to U and V and, upon satisfying condition 2), make the required turn of the axes in U, which would leave condition 2) valid; and for 4) we would require to choose a respective orthogonal transformation, which would not affect the value of Δ. All these transformations, as we noted above, do not affect further considerations.

Under imposed conditions, the isometricity of the metrics (1.1) through the equality

$$g_{ij}(x) = h_{pq}(f(x)) \frac{\partial f^p}{\partial x^i} \frac{\partial f^q}{\partial x^j} \tag{1.31}$$

leads to the following equations:

$$g_{ij} = \delta_{ij} + G_{ij}(x) = \delta_{ij} + H_{ij}(f(x)) + \frac{\partial F^i(x)}{\partial x^j} + \frac{\partial F^j(x)}{\partial x^i} +$$

$$H_{iq}\frac{\partial F^q}{\partial x^j} + H_{pj}\frac{\partial F^p}{\partial x^i} + \delta pq\frac{\partial F^p}{\partial x^i}\frac{\partial F^q}{\partial x^j} + H_{pq}\frac{\partial F^p}{\partial x^i}\frac{\partial F^q}{\partial x^j},$$

from which we have

$$\frac{\partial F^i}{\partial x^j} + \frac{\partial F^j}{\partial x^i} = Q_{ij}(x) + A_{ij}(x), \tag{1.32}$$

where $Q_{ij} = G_{ij}(x) - H_{ij}(f(x)) \in C^{0,\alpha/2}, |Q_{ij}(x)| \leq C|x|^\alpha$, and A_{ij} are finite sums of the form

$$\sum_{p,q} B_{pq}(x)C_{pq}(x)$$

with $B_{pq}(x), C_{pq}(x) \in C^{0,\alpha/2}, |B_{pq}(x)|, |C_{pq}(x)| \leq C|x|^\alpha$.

Let us single out from (1.32) the equations, in which $i = 1$ or $i = j > 1$. We have

$$\frac{\partial F^i}{\partial x^i} = \frac{1}{2}(Q_{ii} + A_{ii}), i \geq 1$$

$$\frac{\partial F^j}{\partial x^1} + \frac{\partial F^1}{\partial x^j} = Q_{1j} + A_{1j}, \ j > 1. \tag{1.33}$$

Let us fix some $j > 1$ and consider the restrictions of the functions F^1 and F^j to the plane $x^1 O x^j$. By letting $z = x^1 + ix^j$, $i^2 = -1$, and $w(z) = F^1(z) + iF^j(z)$, $F^k(z) = F^k(x^1, 0, \ldots, 0, x^j, 0, \ldots, 0)$, $k = 1$ or j, from (1.33) we obtain equations of the form:

$$\frac{\partial w}{\partial \bar{z}} = p(z), \ Re\frac{\partial w}{\partial z} = q(z); \tag{1.34}$$

in the right-hand side the function $p(z)$ satisfies the conditions of Lemma 1.2. Therefore, $|w(z)| \leq C|z|^{1+\alpha}$; in particular, for $z = b^1 + 0i$ we get

$$|F^1(b^1, 0, ..., 0) + iF^j(b^1, 0, ..., 0)| \leq C|b - a|^{1+\alpha}$$

or

$$|F^1(b)| \leq C|b - a|^{1+\alpha}, \ |F^j(b)| \leq C|b - a|^{1+\alpha}, a = 0, j > 1.$$

As $\Delta^2 = \sum_{j=1}^{n}(F^j(b))^2$, we have the estimate of the form

$$|f(b) - f(a) - f'(a)(b - a)| \leq C_0|b - a|^{1+\alpha}.$$

In the chosen coordinates, we have

$$|f(b) - f(a) - f'(a)(b - a)| = |f(b) - f(a) - f'(a)(b - a)|_{h(f(a))},$$

$$|b - a| = |b - a|_{g(a)},$$

so the estimate (1.30) is proved. But since, as we said above, it does not depend on the special choice of the system of coordinates, the estimate of this form is valid for all pairs of points a, b from some sufficiently small ball $\omega : |x| < r_0$. However, inequalities (1.5) are

                I.Kh. SABITOV

also valid in this ball, so, together with (1.30), the estimate of the form

$$|f(b) - f(a) - f'(a)(b-a)| \le C_0|b-a|^{1+\alpha}, \quad \forall a, b : \ |a|, |b| < r_0 \quad (1.35)$$

is valid too (we do not specify the values of the constants participating in the inequalities as coefficients). And with such an inequality present, we are already capable of proving (see the considerations after the inequality (1.9)) that $f \in C^{1,\alpha}$.

Let now the metrics (1.1) belong to $C^{0,1}$. Then $f \in C^{1,\alpha}$ with any $\alpha < 1$; in particular, we can consider that $\alpha > 1/2$. Then the equations of the system (1.34) shall satisfy the conditions of Lemma 1.3, and, thus, instead of the estimate (1.35), we shall have

$$|f(b) - f(a) - f'(a)(b - a)| \le C_0|b - a|^2, \quad \forall a, b \ c \ |a|, |b| < r_0,$$

which will give us that the isometry $f \in C^{1,1}$.

Thus, if ds^2 and $d\sigma^2$ are of class $C^{0,\alpha}, 0 < \alpha \le 1$, then $f \in C^{1,\alpha}$. But if ds^2 and $d\sigma^2$ are of C^1, then we first get that $f \in C^{1,1}$. After that we note that f has Sobolev's second-order derivatives with mixed derivatives not depending on the order of differentiation. In this case, taking into consideration the formula of the change of Christoffel coefficients of a metric in passing to other coordinates, we have

$$\frac{\partial^2 f^m}{\partial x^i \partial x^j} = \Gamma_{ij}^p \frac{\partial f^m}{\partial x^p} - \Delta_{pq}^m \frac{\partial f^p}{\partial x^i} \frac{\partial f^q}{\partial x^j}, \quad (1.36)$$

where Γ_{ij}^p and Δ_{pq}^m are the Christoffel coefficients for, respectively, ds^2 and $d\sigma^2$. Hence, the second derivatives from f^m are in reality continuous, and therefore $f \in C^2$. Successive application of the same formula (1.36) proves the theorem in all classes of $C^{k,\alpha}, 1 \le k \le \infty, \ 0 \le \alpha \le 1$.

It remains to consider the case of analytical metrics. For this, we add up equations (1.36) at each m and all $i = j$. We get that each component f^m of the mapping f satisfies an equation of the form $\Delta f^m = p_m(x, f, f')$, where Δ is the Laplace operator in the n-dimensional space of variables $(x^1, \ldots, x^n)$, and $p_m(x, f, f')$ denotes some function analytical on x and the components of the mapping f and their first-order partial derivatives. Together, all these n equations shall compose a quasilinear elliptic (in the sense of Petrowsky) system relative to n components of $(f^1, \ldots, f^n)$, and it is known that solutions of such a system are analytical functions. Theorem 1.1 is proved.

Remark 1.1. Let us make some remarks to the theorem and to its proof.

1) The starting point of our proof was the Calabi–Hartman theorem affirming the $C^{1,\alpha/2}$ smoothness of the isometry with the metrics (1.1) being of class $C^{0,\alpha}$. In actual fact, we are only interested to know that the isometry is at least of some class $C^{1,\varepsilon}, \varepsilon > 0$. Then the method of our proof yields that the isometry is of class $C^{1,\beta}, \beta = min\{2\varepsilon, \alpha\}$, so, repeating, in the case of $2\varepsilon < \alpha$, our argument, we would gradually arrive at the required statement that $f \in C^{1,\alpha}$.

2) If one or both metrics are only continuous (i.e., of class C), the isometry between them may not be of class C^1. Calabi and Hartman [35] give an example of a continuous metric $ds^2(u, v)$, isometric to the Euclidean metric $dx^2 + dy^2$, the isometry $x = x(u, v)$, $y = y(u, v)$ between which is not of class C^1 (in [135], this example is generalized for any n-dimensional case). This example shows that the smoothness of an isometric immersion of the metric $ds^2 = g_{ij}du^i du^j$ as a solution of the system $\mathbf{r}_i\mathbf{r}_j = g_{ij}$ for the position vector $\mathbf{r}(u^1, \ldots, u^n)$ (in the above mentioned example, the $C^{0,1}$-smooth solution of the system $\mathbf{r}_x^2 = 1, \mathbf{r}_x\mathbf{r}_y = 0, \mathbf{r}_y^2 = 1, \mathbf{r} = (u(x, y), v(x, y), 0))$, and the smoothness of the obtained surface as the image of an isometric immersion (in the given example, the plane $(u, v, 0)$), may, generally speaking, not coincide. This fact is remarked in [197] too.

3) The following remark is contiguous with this observation: if for two isometric surfaces of different smoothnesses $C^{n,\alpha}, n + \alpha > 0$, and $C^{k,\beta}, k + \beta > 0$, the isometry is established by the equality of the intrinsic coordinates (the approach used to prove many theorems), then the regularity of coordinate representation of a smoother surface can be decreased.

4) The formulated theorem on the smoothness of isometries is unimprovable in the following sense: if one of two isometric metrics is of exact class $C^{k,\alpha}$ (i.e., of no class $C^{m,\beta}$ with $m > k$ or $m = k$, but $\beta > \alpha$), and the other has a greater smoothness, the isometry f between them should evidently be of exact class $C^{k+1,\alpha}$. If, however,

both metrics have the exact smoothness $C^{k,\alpha}$, the isometry f may prove even analytical (to see the possibility of such a situation, it is sufficient to take any metric of smoothness $C^{k,\alpha}$ and make the analytical substitution of coordinates for it; then in new coordinates the metric will still be of class $C^{k,\alpha}$, while the isometry between two representations of the same metric will be analytical).•

Remark 1.2. The isometry of Riemannian spaces can be also defined as a bijective mapping preserving the distances between points (where the distance is understood as the infimum of the lengths of curves connecting the considered pair of points). In [65, Chapter 5], with reference to [117] (see [35] for remarks about this work) it is asserted that the isometries between Riemannian spaces in the sense of metric spaces (i.e., of preserving the distances) and in the sense of preserving the metric form (i.e., of preserving the scalar product or, which is the same, of preserving the lengths of the curves) coincide. Now we can say under which minimal conditions the isometry in the sense of metric spaces will have a certain smoothness.•

As in the case of the continuity of the metrics the isometry can fail to be of class C^1, it is natural to put the following:

Open question 1.1. Under which additional conditions the isometry between isometric Riemannian spaces with continuous metrics will be of class C^1? Apparently, it suffices to assume for this that the geodesics in these metrics emanate from each point with definite directions, and the requirement for their contingency of tangents to be located in some angle of less than π can prove to be insufficient for the smoothness of the isometry.•

1.3 Structure of isometries

The formula (1.7) relating two isometric metrics and the isometry between them makes it possible to elucidate the structure of an isometric mapping. Let $y = f(x)$ be the isometry between the metrics (1.1), and let the metrics $ds^2 = g_{ij}(x)dx^i dx^j$ and $d\sigma^2 = h_{ij}(y)dy^i dy^j$ both be reduced in points $x = x_0$ and $y = y_0 = f(x_0)$ by some linear mappings $\tilde{x} = L(x_0)x$ and $\bar{y} = M(y_0)y$ to, respectively, the normal form $ds^2(\tilde{x}) = \tilde{g}_{ij}(\tilde{x})d\tilde{x}^i d\tilde{x}^j$ and $d\sigma^2(\bar{y}) = \bar{h}_{ij}(\bar{y})d\bar{y}^i d\bar{y}^j$ with

$\tilde{g}_{ij}(\tilde{x}_0) = \delta_{ij}$ and $\bar{h}_{ij}(\bar{y}_0) = \delta_{ij}$. Then the relation (1.7) is written down as

$$\tilde{G} = (\bar{f}'_{\tilde{x}})^T \bar{H} \bar{f}'_{\tilde{x}}, \qquad (1.37)$$

where $\tilde{G}(\tilde{x}) = (L^{-1})^T G(L^{-1}\tilde{x})L^{-1}$, $\bar{H}(\bar{f}(\tilde{x})) = (M^{-1})^T H(\bar{f}(\tilde{x}))M^{-1}$, $\bar{f}(\tilde{x}) = Mf\left(L^{-1}(\tilde{x})\right)$. From (1.37) in point x_0 we can see that the matrix $\bar{f}'_{\tilde{x}_0}$ is some orthogonal matrix $\mathcal{O}(\tilde{x}_0)$. Substituting the values of $\tilde{G}$ and $\bar{H}$ in (1.37) and comparing the obtained with (1.7), we have the following representation for $f'(x)$:

$$f'(x) = M^{-1}(f(x))\mathcal{O}(x)L(x), \qquad (1.38)$$

where $L(x)$ and $M(y)$ are the linear-transformation matrices reducing the metrics ds^2 and $d\sigma^2$ to the normal form, respectively, in points x and $y = f(x)$. The matrices $L(x)$ and $M(f(x))$ are algebraically determined by metric coefficients $g_{ij}(x), h_{ij}(y) \in C^{0,\alpha}$, so their elements shall be of class $C^{0,\alpha}$; therefore, the $C^{0,\alpha}$ smoothness of $f'(x)$ totally depends on our ability to show that the orthogonal matrix $\mathcal{O}(x)$ shall be of class $C^{0,\alpha}$. As the orthogonal matrix affects in a sense only the rotations set by the matrix $f'(x)$, it was quite natural in the Calabi–Hartman proof to study the behaviour of the tangents to the geodesics. Along them the tangents are in an intrinsic sense transferred in the parallel way, i.e., without turns, and the subordination of their rotation in the sense of extrinsic Euclidean metric in the space of variables $(x^1, \ldots, x^n)$ to the Hölder condition is what eventually leads to the establishment of the Hölder property of elements of the orthogonal matrix $\mathcal{O}(x)$. (In the above mentioned example from [35] of two isometric continuous metrics, the isometry f between which is not of class C^1, the representation for f' has a disturbed continuity of exactly the multiplier $\mathcal{O}$.) Besides, as we shall see further in a two-dimensional case, knowledge of the structure of orthogonal-matrix elements can help in finding the matrix f'.

1.4   Application of harmonic coordinates

We remind the reader that, in a given Riemannian space, *harmonic* coordinates are those, in which each coordinate component is a harmonic function, i.e., satisfies the Laplace–Beltrami equation with respect to a given metric (see, e.g., [150]). The Laplace–Beltrami equation in the local coordinates $(x) = (x^1, \ldots, x^n)$ has the form

$$g^{ij}(x)\frac{\partial^2}{\partial x^i \partial x^j} - g^{ij}(x)\Gamma^p_{ij}(x)\frac{\partial}{\partial x^p} = 0,$$

where g^{ij} are countervariant components of a metric tensor, and Γ^p_{ij} are Christoffel coefficients of the second kind. Thereby, analytically the harmonicity of the coordinates (x) means that g^{ij} and Γ^p_{ij} are related by the equality

$$g^{ij}\Gamma^p_{ij} = 0, \ p = 1, \ldots, n.$$

By the very definition of harmonic coordinates, it is seen that in these coordinates the metric tensor should have the smoothness of no less than C^1 (in actual fact, it is sufficient to assume the existence of Sobolev derivatives; also acceptable is the regularity class $C^{0,1}$). If the coordinates (x) are not harmonic, then to change over to the harmonic coordinates $(\xi) = (\xi^1, \ldots, \xi^n) = (\xi^1(x), \ldots, \xi^n(x))$ it is sufficient to find n solutions of the equation

$$g^{ij}(x)\frac{\partial^2 \xi^m}{\partial x^i \partial x^j} - g^{ij}(x)\Gamma_{ij}{}^p(x)\frac{\partial \xi^m}{\partial x^p} = 0 \qquad (1.39)$$

with the nondegenerate Jacobian matrix. It is seen that, first, if in the coordinates (x) the metric was of class $C^{k,\alpha}$, $k \geq 1, 0 < \alpha < 1$, then there are harmonic coordinates, and the transition $\xi = \xi(x)$ to them by the known theorems on the regularity of the solutions of elliptic equations has the smoothness of class $C^{k+1,\alpha}$; second, in the considered (i.e., Hölder) classes of smoothness the order of regularity of a metric in the harmonic coordinates (ξ) shall be not lower than it was in the other arbitrary coordinates (x). The third property – that the transition from some harmonic coordinates (ξ) to other harmonic coordinates (η) is given by the solutions of the equation

$$g^{ij}(x)\frac{\partial^2 \xi^m}{\partial x^i \partial x^j} = 0 \qquad (1.40)$$

and is, for this reason, a transformation of smoothness $C^{k+2,\alpha}$ – was also known earlier [150], but that proof implicitly assumed the considered harmonic coordinates to be related *a priori* by a transition of smoothness of at least class C^2. But now we are in a position to show that this is really so: *any* two systems of harmonic coordinates for a given metric or even for two isometric metrics of class $C^{k,\alpha}, k \geq 1, 0 \leq \alpha \leq 1$, are related by an isometry, which, according to Theorem 1.1, shall certainly be of smoothness class $C^{k+1,\alpha}, k+1 \geq 2$. Then this isometry is a C^2-smooth transition, say, from ξ to η, and is thus given by the solutions of an equation of the kind of (1.40) and shall in fact be of class $C^{k+2,\alpha}, \; k \geq 1, 0 < \alpha < 1$. Thereby, we found a *natural* example of isometric metrics, the isometry between which has the smoothness of two orders of magnitude higher than the order of their own smoothness[2]. On the other hand, if the isometric metrics (1.1) have the smoothness of class C^k or $C^{k,1}, k \geq 1$, then Theorem 1.1 guarantees for the isometry between them the smoothness of one order of magnitude higher, and the transition to harmonic coordinates can have the smoothness of only class $C^{k,\beta}$ or, correspondingly, $C^{k+1,\beta}$ with $\forall \beta < 1$ with the respective loss of smoothness of the metric (see an example in [150]). That is to say, in these classes the use of harmonic coordinates can worsen the regularity of the metric, but the smoothness of isometry in these coordinates would still be patently better (of class C^{k+1} or $C^{k+1,1}$, respectively).

Above we used the theorem on the smoothness of isometries to establish the smoothness of transition between *any* two harmonic systems of coordinates. But it may also be the other way round – we can use harmonic coordinates to obtain the smoothness of isometries, in the assumption that metrics have the smoothness of $C^{k,\alpha}, k + \alpha > 1$, and that the isometry is *a priori* of class C^1. Though the proof of this is shorter than the general proof in Section 1.1, it is still rather bothersome, and we do not give it here, as the available Theorem 1.1 is sufficient for our purposes[3]. Still, harmonic coordinates are undoubtedly useful for *constructing* an isometry. Let us see how it is manifested.

---

[2] In actual fact, as we know from Section 1.1, such an isometry may prove even analytical.

[3] Besides, for the case when isometric metrics are of class C^1 and the isometry between them *a priori* belongs to class C^1, the C^2 smoothness of the isometry has been proven in [71] for $n = 2$, and in [72] for any n.

Let the metrics (1.1) be of regularity class $W_k^p, p > n$, functions having Sobolev derivatives of the order of $k \geq 1$ belonging to the class $L_p, p > n$ (we consider this class of regularity for the general scheme to comprise metrics of smoothness $C^{k,\alpha}, \alpha = 0$ or $1, k + \alpha \geq 1$; for classes $C^{k,\alpha}, k \geq 1, 0 < \alpha < 1$, everything proceeds in exactly the same way as for W_k^p). Consider the following diagram:

$$
\begin{array}{ccc}
W_k^p \ni (x) & \xrightarrow{\;\;\varphi \in W_{k+1}^p\;\;} & (\xi) \in W_k^p \\
f \downarrow & & \downarrow g \in W_{k+2}^p \\
W_k^p \ni (y) & \xrightarrow{\;\;\psi \in W_{k+1}^p\;\;} & (\eta) \in W_k^p
\end{array}
$$

The given isometric metrics in their coordinates are denoted in it as (x) and (y) with indication of the smoothness classes of the metric tensor and transitions φ and ψ to some corresponding harmonic coordinates (ξ) and (η), for which the respective smoothness of the metric tensor is also given. The mapping g transfers some harmonic coordinates to others and has the smoothness W_{k+2}^p. All these transitions are obtained as solutions of the known linear equations (1.39) and (1.40). Then the isometry $f : (x) \to (y)$ is obtained as the composition $\psi^{-1} \circ g \circ \varphi \in W_{k+1}^p$. Thus, we obtained the method of *representing* isometries through some standard operations. This representation cannot yet be called *canonical*, as there is great freedom in the choice of all three of its components, and we should be able somehow to standardize this choice. In particular, there is no answer to the following:

Open question 1.2. If an isometry f is already given, how can we choose a composition to match it?•

Transition to harmonic coordinates can also give additional information on the isometries, which is useful for finding the sought-for isometry itself. Namely, let the transition to harmonic coordinates be made in both isometric metrics. We shall consider that the coordinates (x) and (y) in (1.1) are already harmonic. Then let us make up from equations (1.36) a pairwise sum (i, j):

$$
g^{ij} \frac{\partial^2 f^m}{\partial x^i \partial x^j} = g^{ij} \Gamma_{ij}^p \frac{\partial f^m}{\partial x^p} - g^{ij} \Delta_{pq}^m \frac{\partial f^p}{\partial x^i} \frac{\partial f^q}{\partial x^j}.
$$

As x^m and $y^m = f^m$ are harmonic coordinates, then on the strength of (1.39) and (1.40) we have

$$g^{ij}\Gamma^p_{ij}\frac{\partial f^m}{\partial x^p} = 0, \quad g^{ij}\frac{\partial^2 f^m}{\partial x^i \partial x^j} = 0,$$

hence, besides (1.31), we also have n more equations for elements of the matrix f':

$$g^{ij}(x)\Delta^m_{pq}(f(x))\frac{\partial f^p(x)}{\partial x^i}\frac{\partial f^q(x)}{\partial x^j} = 0, \quad m = 1, 2, \ldots, n.$$

Remark 1.3. The existence and smoothness properties of harmonic coordinates have been studied in metric spaces, *a priori* more general than Riemannian spaces with the metric tensor of smoothness C^2 (see [2, 15, 134, 136]).•

2 Two-dimensional locally Euclidean metrics

2.1 *Smoothness of two-dimensional isometries*

In the case of an isometry between metrics on two-dimensional manifolds, the regularity of isometry can be obtained by reducing the problem to the regularity of solutions of elliptic systems. Indeed, let (u, v) and (x, y) be the new notation of the coordinates (x^1, x^2) and (y^1, y^2), respectively. Let us rewrite the relation (1.7) in the form

$$F = H^{-1}(F^{-1})^T G.$$

Substituting here the values

$$F(u, v) = \begin{pmatrix} x_u & x_v \\ y_u & y_v \end{pmatrix} \text{ and } det(F)F^{-1} = \begin{pmatrix} y_v & -x_v \\ -y_u & x_u \end{pmatrix}$$

and taking into account that $det^2(F) = \dfrac{detG}{detH}$, we obtain four equations, from which we can single out the following quasilinear elliptic system:

$$\begin{cases} (g_{22}x_u - g_{12}x_v)\sqrt{\Delta_1} & = (h_{21}x_v + h_{22}y_v)\sqrt{\Delta_2} \\ (g_{21}x_u - g_{11}x_v)\sqrt{\Delta_1} & = (h_{21}x_u + h_{22}y_u)\sqrt{\Delta_2} \end{cases} ; \qquad (2.1)$$

26   I.Kh. SABITOV

here it is set that $\Delta_1 = \det H(x(u,v), y(u,v))$, $\Delta_2 = \det G(u,v)$ (the *a priori* existence of the derivatives x_u etc. is guaranteed in at least a generalized sense since the isometry $f \in C^{0,1}$). Let the metrics (1.1) have the regularity of class $C^{k,\alpha}$, $k \geq 0, 0 < \alpha < 1$. First assume that $k = 0$. Then the coefficients of the system (2.1) as a function of u, v have the smoothness $C^{0,\alpha}$ and, by the known theorems on the regularity of elliptic systems' solutions, we obtain that $x(u,v)$, $y(u,v) \in C^{1,\alpha}$. If $k = 1$, then, establishing first that $f \in C^{1,\alpha}$, we get that the coefficients of the system will also be of class $C^{1,\alpha}$, so its solution $f \in C^{2,\alpha}$. At $k > 1$, repeating this reasoning k times, we arrive at the conclusion that $f \in C^{k+1,\alpha}$. However, these theorems do not work for the cases $\alpha = 0$ and $\alpha = 1$, so then we need to refer to the above given proof of Theorem 1.1 and to use it as the justification of, respectively, $C^{k+1,0}$ or $C^{k+1,1}$ smoothness of solutions of the elliptic system (2.1) at the smoothness coefficients $C^{k,0}$ and $C^{k,1}$.

For two-dimensional metrics of class C^2 and higher a particular case of harmonic coordinates are *isothermal coordinates*, in which the metric has the form $ds^2 = \Lambda^2(w)(du^2 + dv^2), w = u + iv$. But isothermal coordinates also exist for metrics of lower smoothness[4], and transition to them is given by the solution of the Beltrami system (transition from arbitrary coordinates (x, y) to isothermal coordinates (u, v) is assumed):

$$\frac{\partial w}{\partial \bar{z}} = q(z) \frac{\partial w}{\partial z}, \quad z = x + iy, \quad w = u + iv, \qquad (2.2)$$

where $q(z) = \dfrac{g_{11} - g_{22} + 2ig_{12}}{g_{11} + g_{22} + 2\sqrt{\Delta}}, \Delta = g_{11}g_{22} - g_{12}^2$. Also valid here are the theorems on the increase of smoothness of the solution by one unit as compared with the smoothness of the coefficient (in Hölder classes $C^{k,\alpha}, 0 < \alpha < 1$, or in Sobolev classes $W_k^p, p > 2$). For the results on these and other properties of Beltrami system solutions, see [16, 184] and recent works [21, 67, 109, 199]. If the coefficients of the metric form $ds^2(x, y)$ are of classes $C^{k,0}$ or $C^{k,1}$, and if it is known beforehand that the coefficient Λ in the isothermal form is also of the same class of smoothness (which is just the case of l. E. metrics),

---

[4] Even for manifolds of bounded curvature in the Alexandrov sense without any prior requirements of smoothness, see [134, 136].

then the corresponding one-unit-larger smoothness of transition to isothermal coordinates (or, which is the same, the smoothness of the isometry between two forms of a metric) is obtained not from the theory of elliptic equations, but from Theorem 1.1. This observation makes the statement of the following problem natural.

Open question 2.1. Let the coefficients of some elliptic system of first-order equations for n unknown functions be of class $C^{k,1}$ or $C^{k+1,0}$, $k \geq 0$, and let it be known that the Jacobian of the system of these functions is not equal to zero and is of the same class of smoothness as the coefficients of the equations. Can it be asserted that the smoothness of the solution of the system is one unit higher than the smoothness of the coefficients? The same question with a respective change of formulation can be also put for highest-order elliptic systems.$\bullet$

It is now natural to ask if Theorem 1.1 can also be proved for metrics of class $C^{0,\alpha}, \alpha > 0$, in the cases when $n \geq 3$ too, by reducing the problem to the smoothness of the solutions of elliptic systems? The difficulty is that in these cases we obtain $N = \dfrac{n(n+1)}{2}$ first-order nonlinear equations for n unknown functions, from which we fail to single out an elliptic combination without additional differentiation. However, a stronger assumption of the smoothness of class C^1 in [72] and [81] led us to show how to pass from first-order systems to second-order equations (similar to the system (1.36)) and to obtain the required proposition on the smoothness of isometry. Thus, we have:

Open question 2.2. How to find a *new* and shorter proof of the initial smoothness of isometry (i.e., their smoothness of at least class $C^{1,\varepsilon}$ at some $\varepsilon > 0$)?$\bullet$

2.2 *Characteristic features and isometries of locally Euclidean metrics*

2.2.1 *Statement of the problem*

The general problem is formulated thus: if a metric is given in local coordinates by some quadratic form $ds^2 = g_{ij}(x)dx^i dx^j$, then how can we determine by its coefficients whether this metric is locally

Euclidean or not? When we are speaking of the "determination by its coefficients", we mean that we need to indicate some actions to be done with the coefficients, so that as the result it could be said whether this metric is locally Euclidean or not. Apparently, it is rather difficult to formulate a clear definition of what can admissibly be called an "action" over the coefficients. In the arbitrary understanding of the list of "actions", the check for local Euclidity can be done for a very broad range of metric spaces. For this, for instance, the concept of curvature in the Alexandrov sense is introduced, which assumes knowledge of shortest lines and skills of calculating the angles in triangles with shortest sides so that it could be verified if the excess of the sum of angles in any triangle is equal to zero or not, see [3]. In the two-dimensional case, there is an analytical approach for solving this problem: it is necessary to pass to isothermal coordinates and to check whether the logarithm of metric's coefficient in these coordinates is a harmonic function or not, see [134, 136]. The second test will be discussed below; as for the first test, we can at once put such a question: how can shortest lines be found if at a small smoothness of metric's coefficients we cannot even make up differential equations of the geodesics? And finding the shortest lines by the Hilbert method mentioned on p. 5 requires, as a minimum, knowledge of the lengths of the shortest lines, which, in fact, we do not know how to calculate. In the meantime, in our approach to the problem, we would like to perform only those operations over the coefficients, which would assume the calculation of a *finite number* of some algebro-functional, differential and integral expressions or the solution of a finite number of algebro-functional, differential and integral equations, preferably with the known methods of finding their solution with arbitrarily high accuracy.

If the coefficients of the metric are sufficiently smooth, namely, of class C^2 or higher, such a test for local Euclidity is known. For this, it is necessary and sufficient to verify if the metric's curvature tensor calculated using second derivatives of the metric's coefficients is equal to zero or not. In the next Section we examine in detail the case of smoothness C^1 for two-dimensional metrics, simultaneously obtaining the formulae for finding the isometry proper for the standard Euclidean metric. Further on, we shall mainly abide by our work [145].

2.2.2   Isometries of two-dimensional locally Euclidean metrics in quadratures

Let an l. E. metric

$$ds^2 = E(u, v)du^2 + 2F(u, v)dudv + G(u, v)dv^2 \qquad (2.3)$$

be given in some domain D of the plane of variables (u, v), and we shall not yet make the smoothness of the coefficients more precise. Let

$$x = x(u, v), y = y(u, v) \qquad (2.4)$$

be an isometry between the metric (2.3) and the standard metric $d\sigma^2 = dx^2 + dy^2$. By the definition of isometry, we have

$$x_u^2 + y_u^2 = E, \;\; x_u x_v + y_u y_v = F, \;\; y_u^2 + y_v^2 = G,$$

whence it follows that there are some functions α and β, such that $x_u = \sqrt{E}\cos\alpha$, $y_u = \sqrt{E}\sin\alpha$, $x_v = \sqrt{G}\cos\beta$, $y_v = \sqrt{G}\sin\beta$. Setting $\alpha + \beta = 2\varphi$, $\alpha - \beta = 2\theta$, we arrive at the formulae

$$x_u = \sqrt{E}\cos(\varphi + \theta), \;\; x_v = \sqrt{G}\sin(\varphi - \theta),$$

$$(2.5)$$

$$y_u = \sqrt{E}\sin(\varphi + \theta), \;\; y_v = \sqrt{G}\cos(\varphi - \theta).$$

Using the equality $x_u x_v + y_u y_v = F$, we get that $\sin 2\varphi = \dfrac{F}{\sqrt{EG}}$. In the assumption that the isometry preserves orientation, we also have the inequality $\cos 2\varphi > 0$, so if we additionally require that the condition $|2\varphi| < \pi/2$ be satisfied in one point, the value φ will be uniquely determined by the following formulae:

$$\cos\varphi = \frac{\sqrt{\delta + \Delta}}{\sqrt{2\delta}}, \sin\varphi = \frac{F}{\sqrt{2\delta(\delta + \Delta)}}, \delta = \sqrt{EG}, \Delta = \sqrt{EG - F^2}.$$

$$(2.6)$$

Note that the obtained formulae (2.5) for the isometric mapping f exactly correspond to their theoretical form (1.38):

$$f'(u, v) = \begin{pmatrix} x_u & x_v \\ y_u & y_v \end{pmatrix} =$$

$$\mathcal{O}L^{-1} = \begin{pmatrix} \cos\theta & -\sin\theta \\ \sin\theta & \cos\theta \end{pmatrix} \begin{pmatrix} \sqrt{E}\cos\varphi & \sqrt{G}\sin\varphi \\ \sqrt{E}\sin\varphi & \sqrt{G}\cos\varphi \end{pmatrix}$$

(in our case the matrix M is equal to the unit matrix). The Jacobian of the mapping (2.4) is equal to $J = \Delta = \delta\cos 2\varphi$.

The compatibility condition of each pair of equations of the system (2.5) leads to a new system, but this time for the determination of θ:

$$\theta_u = \varphi_u + \frac{F}{2\Delta}(\ln G)_u - \frac{E_v}{2\Delta}, \quad \theta_v = -\varphi_v - \frac{F}{2\Delta}(\ln E)_v + \frac{G_u}{2\Delta}. \quad (2.7)$$

Thus, for the local Euclidity of the metric (2.3) it is necessary and sufficient that for the system (2.7) the condition of its compatibility be satisfied, which, given that the metric is of class C^2, leads exactly to the requirement of equality of the Gaussian curvature K of the metric (2.3) to zero. This is the known classical condition of the local Euclidity of a metric.

We get that, in the classical assumptions of smoothness, the isometric immersion of the metric (2.3) into E^2 requires the solution of two problems in quadratures: first from (2.7) we find the function $\theta(u, v)$ by its derivatives θ_u and θ_v, then by the system (2.5) we find $x(u, v)$ and $y(u, v)$, also by their partial derivatives. On the strength of Theorem 1.1, the isometry shall be of class $C^{m+1,\alpha}$, where $C^{m,\alpha}, m + \alpha \geq 1$, is the smoothness of the metric; therefore, this method of solving the problem is possible based on the assumption that the coefficients of the metric are of smoothness class $C^{0,1}$ and higher.

What has been said is summed up in the following theorem.

Theorem 2.1. *Let the coefficients of the l. E. metric (2.3) be of class $C^{m,\alpha}$, $m \geq 0$, $0 \leq \alpha \leq 1$, $m + \alpha > 0$ in some domain $\bar{D}$. Then each interior point of the domain has a neighbourhood, which can be isometrically embedded into E^2 by a mapping of smoothness $C^{m+1,\alpha}$. In an additional assumption that $m + \alpha \geq 1$, this embedding can be found in quadratures; if, in this case, the domain D has the boundary of smoothness $C^{m'+1,\alpha}$, $m' \geq 0$, the entire closed domain is immersed into E^2, and this immersion transfers the boundary of the domain D into a curve of smoothness $C^{p+1,\alpha}$, where $p = min(m, m')$.*

Remark 2.1. The fact that an isometric immersion has the smoothness one order higher than the smoothness of the coefficients of a given metric follows, as we have already mentioned above, from the theorem on the smoothness of isometries in the previous Section. However, this theorem gives no information on the smoothness of isometry at the boundary, so to obtain a proposition on the smoothness of isometry in $\bar{D}$, it is necessary to use the explicit form of the systems (2.5) and (2.7). But the system (2.7) is written down only for metrics of at least class $C^{0,1}$, so in this part of the theorem we had to assume additionally that metrics of class $C^{m,\alpha}$ with $m+\alpha \geq 1$ are being considered.

In the case when the domain of existence of the metric (2.3) is a circle $\Omega : u^2 + v^2 \leq R^2$ (and the case of an arbitrary simply connected domain D can be reduced to the case of a circle by preliminary conformal mapping), we can propose explicit formulae giving the criterion of a local Euclidity of the metric (2.3) in cases of smoothness $C^{0,1}$ and higher, and also making it possible to present an isometric immersion into E^2 in a compact and calculable form. For this, we prove a lemma on the presentation of a solution to the problem of finding a real function $F(u, v)$ by its partial derivatives in the form of some double integral operator.

Lemma 2.1. *Let the equation* $\dfrac{\partial F}{\partial \bar{w}} = H(u, v)$ *be given in a circle* $\bar{\Omega}$,

where $\dfrac{\partial}{\partial \bar{w}} = \dfrac{1}{2}\left(\dfrac{\partial}{\partial u} + i\dfrac{\partial}{\partial v} \right)$, $w = u + iv \in \bar{\Omega}$, *and the right-hand side is of class* $C^{m,\alpha}$, $m \geq 0, 0 \leq \alpha \leq 1$. *Then for the solvability of this equation in the class of real single-valued functions of smoothness* $C^{m+1,\alpha}$, *it is necessary and sufficient that the function* $H(u, v)$ *satisfy the condition*

$$Im\left(\frac{1}{\pi} \iint\limits_{\Omega} \frac{R^2 - |w|^2}{(\zeta - w)(R^2 - \bar{w}\zeta)} H(\xi, \eta)\, d\xi d\eta \right) = 0, \quad \zeta = \xi + i\eta \in \bar{\Omega}.$$

$$(2.8)$$

In this case, the sought-for real function $F(u, v)$ *with the condition* $F(0, 0) = 0$ *is determined by the formula*

$$F(u, v) = Re\left[-\frac{1}{\pi}\iint\limits_{\Omega} \frac{wR^2 - \bar{w}\zeta^2}{\zeta(\zeta - w)(R^2 - \bar{w}\zeta)}H(\xi, \eta)d\xi d\eta\right]. \quad (2.9)$$

Proof. The general solution of the equation $\dfrac{\partial F}{\partial \bar{w}} = H(u, v)$, as we already know, has the form (see [184], Chapter 1)

$$F(u, v) = -\frac{1}{\pi}\iint\limits_{\Omega} \frac{H(\xi, \eta)}{\zeta - w}\, d\xi d\eta + \Phi^+(w),$$

$$\zeta = \xi + i\eta \text{ and } w = u + iv\ \bar{\Omega}, \quad (2.10)$$

where $\Phi^+(w)$ is a function holomorphic in Ω. Let it first be known that these solutions include the real function $F_0(u, v)$. Consider (2.10) for $F_0(u, v)$ in points of the circumference $\partial\Omega$:

$$F_0(t) = \Psi^-(t) + \Phi^+(t),\ |t| = R, \quad (2.11)$$

where $\Psi^-(t)$ are the limit values on $\partial\Omega$ of the function

$$\Psi^-(w) =$$

$$-\frac{1}{\pi}\iint\limits_{\Omega} \frac{H(\xi, \eta)}{\zeta - w}\, d\xi d\eta \equiv T_\Omega H,\ |w| > R,\ |\zeta| \le R;\ \Psi^-(\infty) = 0$$

holomorphic in the exterior of the circle Ω (in (2.11) it is taken into account that the function $T_\Omega H$ is continuous on the entire plane w, see Chapter 1 in [184]). By assumption,

$$Im F_0(t) = \Psi^-(t) - \overline{\Psi^-(t)} + \Phi^+(t) - \overline{\Phi^+(t)} = 0,\ |t| = R. \quad (2.12)$$

Consider the holomorphic functions

$$\Psi^+(w) = \overline{\Psi^-\left(\frac{R^2}{\bar{w}}\right)},\ |w| \le R;\ \Phi^-(w) = \overline{\Phi^+\left(\frac{R^2}{\bar{w}}\right)},\ |w| \ge R.$$

As $\Psi^+(t) = \overline{\Psi^-(t)},\ \Phi^-(t) = \overline{\Phi^+(t)}$, from (2.12) we get

$$\Psi^+(t) - \Phi^+(t) = \Psi^-(t) - \Phi^-(t),$$

so

$$\Phi^+(w) - \Psi^+(w) = const = \Phi^-(\infty) - \Psi^-(\infty) = \overline{\Phi^+(0)}.$$

Besides, as $\Psi^+(0) = 0$, $\Phi^+(0) = \bar{\Phi}^+(0)$, i.e., $c = \Phi^+(0)$ is a real number. Therefore,

$$F_0(u,v) = -\frac{1}{\pi} \iint\limits_\Omega \frac{H(\xi,\eta)}{\zeta - w}\, d\xi d\eta$$

$$+ \frac{w}{\pi} \iint\limits_\Omega \frac{\overline{H(\xi,\eta)}}{R^2 - \bar\zeta w}\, d\xi d\eta + c, \, |w| \le R. \tag{2.13}$$

As $F_0(u,v)$ is real not only at the boundary, but also within the entire circle Ω, it is necessary that

$$ImF_0(u,v) = -\frac{1}{2\pi i} \iint\limits_\Omega \left(\frac{1}{\zeta - w} + \frac{\bar w}{R^2 - \zeta \bar w} \right) H(\xi,\eta)\, d\xi d\eta +$$

$$\frac{1}{2\pi i} \iint\limits_\Omega \left(\frac{1}{\bar\zeta - \bar w} + \frac{w}{R^2 - \bar\zeta w} \right) \overline{H(\xi,\eta)}\, d\xi d\eta = 0,$$

$$\tag{2.14}$$

which, after simplification, leads to the condition (2.8).

Subtracting from (2.13) the value $F_0(0,0)$, we obtain a new function $F(u,v)$ equal to zero at $w = 0$, and with provision for the equality $F = ReF$ from (2.13) get for it the representation (2.9).

Let it now be known that $H(u,v)$ satisfies (2.8). Consider the function $F_0(u,v)$ from (2.13) with a real summand c. As (2.8) is equivalent to (2.14), (2.13) determines the real function F_0, which after the subtraction of $F_0(0,0)$ from it yields the function $F(u,v)$ from (2.9) with $F(0,0) = 0$ satisfying the equation $\dfrac{\partial F}{\partial \bar w} = H(u,v)$.

If there is one more real solution $F_1(u,v)$ with $F_1(0,0) = 0$, then on the strength of the equality $\dfrac{\partial}{\partial \bar w}(F - F_1) = 0$ we have that the real function $F_1 - F$ is holomorphic, which is possible only at $F_1 \equiv F$.

Further, for the real function F we have $\dfrac{\partial F}{\partial w} = \dfrac{1}{2}(F_u - iF_v) = \bar H$, so the smoothness F shall be of class $C^{m+1,\alpha}$. The lemma is proved.

We use this lemma to obtain in an explicit form the conditions of the local Euclidity of the metric (2.3) and the formula for its immersion into E^2 given that the smoothness of the metric is not lower than $C^{0,1}$. Using the formulae (2.6), we compute the derivatives φ_u and φ_v and find in (2.7) its right-hand-side parts. We get

$$\theta_u = \frac{(EG_u - GE_u)F + 2EG(F_u - E_v)}{4\Delta\delta^2} \equiv g \qquad (2.15)$$

$$\theta_v = \frac{(EG_v - GE_v)F + 2EG(G_u - F_v)}{4\Delta\delta^2} \equiv h, \qquad (2.16)$$

where $\delta = \sqrt{EG}$, $\Delta = \sqrt{EG - F^2}$. Then we have

$$\frac{\partial\theta}{\partial\bar{w}} = \frac{1}{2}(g + ih) \equiv H_0. \qquad (2.17)$$

Using Lemma 2.1, we obtain that the condition of the local Euclidity of the metric is given by the integral equality

$$Im\left(\frac{1}{\pi}\iint_\Omega \frac{R^2 - |w|^2}{(\zeta - w)(R^2 - \bar{w}\zeta)}H_0(\xi,\eta)d\xi d\eta\right) = 0,\ \zeta = \xi + i\eta \in \bar{\Omega},$$

$$(2.18)$$

and the function $\theta(u,v)$ proper has the following representation:

$$\theta(u,v) = Re\left(-\frac{1}{2\pi}\iint_\Omega \frac{wR^2 - \bar{w}z^2}{\zeta(\zeta - w)(R^2 - \bar{w}\zeta)}H_0(\xi,\eta)d\xi d\eta\right), \quad (2.19)$$

where the function H_0 is determined in the equality (2.17). In turn, writing down the equalities (2.5) as the equations for $\dfrac{\partial x}{\partial\bar{w}}$ and $\dfrac{\partial y}{\partial\bar{w}}$ and again using Lemma 2.1, for the isometric mapping $f = x + iy$ after some transformations we have the following representation:

$$f(w) = x(w) + iy(w) =$$

$$\frac{1}{2\sqrt{2}\pi}\iint_\Omega \sqrt{E}\frac{\Delta + \delta + iF}{\sqrt{\delta^2 + \delta\Delta}}e^{i\theta(\zeta)}Re\frac{\bar{w}z^2 - wR^2}{\zeta(\zeta - w)(R^2 - \zeta\bar{w})}d\xi d\eta -$$

$$\frac{i}{2\sqrt{2}\pi}\iint_\Omega \sqrt{G}\frac{\Delta + \delta - iF}{\sqrt{\delta^2 + \delta\Delta}}e^{i\theta(\zeta)}Im\frac{\bar{w}\zeta^2 - wR^2}{\zeta(\zeta - w)(R^2 - \zeta\bar{w})}d\xi d\eta, \qquad (2.20)$$

where the function θ is determined by the formula (2.19).

Thus, from this simple Lemma 2.1 we derive the following corollaries. First, we get in a compact and algorithmically calculable form the criterion (2.18) of the local Euclidity of the metric (2.3) given that its coefficients are C^1- or $C^{0,1}$-smooth; second, by finding from the system (2.7) its solution (2.19), we find by the formula (2.20) the isometry proper of the metric (2.3) onto the Euclidean plane (x, y). Let us sum up these propositions in the form of the following theorem.

Theorem 2.2. *The simply connected domain $\bar{D}$ with the l. E. metric (2.3) under conditions of the smoothness $E, F, G \in C^{m,\alpha}(\bar{D})$ and $\partial D \in C^{m'+1,\alpha}, m \geq 0, 0 \leq \alpha \leq 1, m + \alpha \geq 1, 0 \leq m' \leq m$, is isometrically immersed into E^2 by a mapping of smoothness $C^{m+1,\alpha}(D)$ with preservation of the smoothness of the boundary[5], which in the case of a circle admits an explicit representation in the form of (2.20).*

We complete this Section by several remarks.

Remark 2.2. The formula (2.20) for reducing the l. E. metric (2.3) to the standard form also remains valid for metrics of class $C^{0,\alpha}$, but now we do not know the value of the function θ in it, because the formula (2.19) cannot be used anymore. Finding the function θ is now reduced to the requirement that it satisfy two conditions of the form of (2.8) from Lemma 2.1, arising from the requirement of the consistency of the equations in each system for $\dfrac{\partial x}{\partial \bar{w}}$ and $\dfrac{\partial y}{\partial \bar{w}}$. We write out these conditions, combining them into one complex equality, which comprises simultaneously both the conditions of the local Euclidity of the metric and the value of the function $\theta(w)$ in the following sense: *for the metric (2.3) with the coefficients of class $C^{0,\alpha}, 0 < \alpha < 1$, given in the circle $\Omega : |w| \leq R$, to be locally Euclidean, it is necessary and sufficient that there exist the real function $\theta(w)$ of class $C^{0,\alpha}$, satisfying the equation*

---

[5] Note that the image of the mapping shall, however, be an *analytical* surface – a domain on the plane.

$$\iint\limits_{\Omega} \sqrt{G}\frac{\Delta + \delta - iF}{\sqrt{\delta^2 + \delta\Delta}}e^{i\theta(\zeta)}Re\frac{R^2 - |w|^2}{(\zeta - w)(R^2 - \zeta\bar{w})}d\xi d\eta - $$

$$-i\iint\limits_{\Omega} \sqrt{E}\frac{\Delta + \delta + iF}{\sqrt{\delta^2 + \delta\Delta}}e^{i\theta(\zeta)}Im\frac{R^2 - |w|^2}{(\zeta - w)(R^2 - \zeta\bar{w})}d\xi d\eta = 0.\bullet \quad (2.21)$$

Remark 2.3. In principle, the proposed formulae for verifying the local Euclidity of the metric (2.3) and finding its isometric mapping in E^2 are not new – as a matter of fact, they are given in [78] and even in [46]. But there they are presented, first, in a non-symmetric form with respect to the coefficients E, F, G; second, there the condition of local Euclidity is formulated as follows: for the local Euclidity of a metric, it is necessary and sufficient that the integral $\oint gdu + hdv$ (for the definitions of g and h, see the formulae (2.15) and (2.16)), be equal to zero *along all* smooth closed curves. In our review here, in accordance with the approach outlined at the beginning of this paragraph, we replace this condition by the requirement of the satisfaction of *one* integral equality (2.18). It is clear that in such a form this condition can be computationally verified (certainly, to a given accuracy) much simpler than the equality of curvilinear integrals to zero for all curves.$\bullet$

Remark 2.4. The field of application of the formulae (2.19) and, respectively, (2.20) can be expanded a little bit by assuming for the metric (2.3) only the existence of the derivatives E_v and G_u. Indeed, using for the computation of θ by Lemma 2.2 the formula (2.7), we can see that by integrating by parts we can reduce the integrals of φ_u and φ_v to the integrals of $\varphi = \arctan\dfrac{F}{\delta + \Delta}$, and for calculating θ we would need to know only the derivatives E_v and G_u.$\bullet$

Remark 2.5. Given in an open simply connected domain, the l. E. metric (2.3) of smoothness C^1 and higher also admits an isometric immersion into E^2 of smoothnesses unity larger than the smoothness of the metric, since it can be represented by corresponding curvilinear integrals within the entire domain of the metric.$\bullet$

2.2.3   *Particular forms of locally Euclidean metrics*

Certainly, in some particular cases the local Euclidity of the metric
can also be verified and can be explicitly reduced to the standard
form without *a priori* assuming the existence of the first derivatives
of its coefficients. Consider some examples.

a) *The metric is given in an isothermal form* with $E = G = \Lambda^2(u, v)$, $F = 0$. In this case, for the local Euclidity of the given metric, it is necessary and sufficient that the function $\ln \Lambda$ be harmonic
[134]. Therefore, to reduce the metric to the standard form, we can
use the formula $(2.20)^6$. In Section 2.3 below, we consider such l. E.
metrics in more detail.

b) *The metric is given in Chebyshev coordinates*, i.e.,

$$ds^2 = du^2 + 2\sin \omega(u, v)dudv + dv^2. \tag{2.22}$$

It is easy to show that if the function ω has the form

$$\omega(u, v) = g(u) + h(v), \tag{2.23}$$

where g and h are continuous functions, the metric (2.22) is locally
Euclidean. Indeed, in the formulas (2.5)–(2.6) we then have $\varphi = g(u)/2 + h(v)/2$ and, by setting $\theta = g(u)/2 - h(v)/2$, we will obtain
the following solution of the system (2.5):

$$x(u, v) = \int \cos g(u)du + \int \sin h(v)dv,$$

$$y(u, v) = \int \sin y(u)du + \int \cos h(v)dv$$

(the necessary changes in these formulae for the standard notation
$\cos \omega$ instead of $\sin \omega$ we used are evident).

Open question 2.3. The necessity of representing (2.23) with the
continuous function ω for the local Euclidity of the metric (2.22)
is unknown (in the assumption of $\omega \in C^1$ the necessity follows from
Theorem 2.1 and the formula (2.7)).●

---

[6] In reality, for this criterion to be valid, we do not require *a priori* even the
continuity of the coefficients of the metric. It is sufficient only that the function
$\Lambda(u, v)$ be positive and measurable along each rectifiable curve in the domain D
of the change of the variables (u, v).

Of the literature on the subject, we can mention [11], which shows that on any manifold of bounded curvature we can introduce the Chebyshev coordinates (u, v), in which the angle $\omega(u, v)$ is a function of bounded variation. Still, it has not been proved there that this property is realized for *any* Chebyshev coordinates.

c) *The metric is given in geodesic polar coordinates*, i.e.,

$$ds^2 = du^2 + G(u, v)dv^2. \tag{2.24}$$

Let us show that if the continuous coefficient $G(u, v)$ is of the form

$$G(u, v) = (g(v) + uh(v))^2, \tag{2.25}$$

the metric (2.24) is reduced to the standard form by quadratures. Indeed, in this case in the formulae (2.5)–(2.6) we have $\varphi = 0$, and if the function θ is chosen as

$$\theta = \int h(v)dv,$$

the system (2.5) has the solution

$$x = u \cos \theta(v) - \int g(v) \sin \theta(v)dv, \quad y = u \sin \theta(v) + \int g(v) \cos \theta(v)dv.$$

Open question 2.4. We are not aware of the proof of the *necessity* of representing $G(u, v)$ in the form of (2.25) for the local Euclidity of the metric (2.24) only in the *a priori* continuity of $G(u, v)$ (in the assumption of $G \in C^1$ the necessity again follows from Theorem 2.1 and equation (2.7)).$\bullet$

d) *The metric is given as the metric of a ruled surface:*

$$ds^2 = (1 + 2(\mathbf{r}'\mathbf{l}')v + \mathbf{l}'^2 v^2)du^2 + 2(\mathbf{l}\mathbf{r}')dudv + dv^2, \tag{2.26}$$

where the vector functions $\mathbf{r} = \mathbf{r}(u) \in C^1, \mathbf{l} = \mathbf{l}(u) \in C^1$; it is additionally known that the vectors $\mathbf{r}'(u)$ and $\mathbf{l}(u)$ are unit vectors and the mixed product

$$(\mathbf{r}', \mathbf{l}, \mathbf{l}') = 0. \tag{2.27}$$

If the condition (2.27) is satisfied, this metric is the metric of a developable ruled surface $S : \mathbf{r}(u) + v\mathbf{l}(u) \in C^1$; however, this is proved by direct formal calculations only in the assumption of $\mathbf{r}(u) \in C^2$. Let us show without turning to the surface S that the metric (2.26) is, under the above mentioned minimal conditions, locally Euclidean when and only when the condition (2.27) is satisfied, and let us simultaneously give the method of reducing it to the standard form.

First we prove the sufficiency of the condition (2.27). We solve a local problem and so can assume one of the following possibilities to be satisfied: either the vectors $\mathbf{l}(u)$ and $\mathbf{r}'(u)$ are nowhere parallel in a sufficiently small neighbourhood of the considered point (u_0, v_0) or the vector $\mathbf{l}(u_0)$ is parallel to $\mathbf{r}'(u_0)$.

In the former case, on the strength of (2.27) the vector $\mathbf{l}'$ can be decomposed relative to the vectors $\mathbf{l}(u)$ and $\mathbf{r}'(u)$:

$$\mathbf{l}'(u) = a(u)\mathbf{l} + b(u)\mathbf{r}'.$$

Hence, multiplying this equality in turns by $\mathbf{l}$, $\mathbf{r}'$ and $\mathbf{l}'$, we have

$$a(u) = -b(u)(\mathbf{lr}') \Rightarrow (\mathbf{l}'\mathbf{r}') = b(u)(1 - (\mathbf{lr}')^2) \Rightarrow \mathbf{l}'^2 = b^2(1 - (\mathbf{lr}')^2).$$

Substituting the found values into (2.26), we obtain:

$$ds^2 = \left(1 - (\mathbf{lr}')^2\right)(1 + vb(u))^2 \, du^2 + ((\mathbf{lr}')du + dv)^2.$$

Choosing the new variables

$$\xi = \int \sqrt{1 - (\mathbf{lr}')^2} du, \quad \eta = v + \int (\mathbf{lr}')du \equiv v + f(u),$$

we arrive at the metric $ds^2 = d\eta^2 + (h(u) + \eta b(u))^2 d\xi^2$, where $h(u) = 1 - f(u)b(u)$, $u = u(\xi)$, i.e., to the form of the l. E. metric considered in the previous Section.

Let now $\mathbf{l}(u_0) \parallel \mathbf{r}'(u_0)$. Then from the nondegeneracy condition of the metric (2.26) and the orthogonality of the vectors $\mathbf{l}(u_0)$ and $\mathbf{l}'(u_0)$ we have that $v_0 \neq 0$, the vectors $\mathbf{l}'(u_0)$ and $\mathbf{r}'(u_0)$ are nonparallel and $(\mathbf{l}'(u_0)\mathbf{r}'(u_0)) = 0$. Then the vector $\mathbf{l}(u)$ is decomposed in the basis $\mathbf{l}'(u), \mathbf{r}'(u)$:

$$\mathbf{l}(u) = \alpha(u)\mathbf{l}' + \beta(u)\mathbf{r}', \quad \alpha(u_0) = 0, \beta(u_0) = \pm 1.$$

From this equation, multiplying it in turns by $\mathbf{l}, \mathbf{l}', \mathbf{r}'$ and substituting the found values of $(\mathbf{r}'\mathbf{l}')$, $(\mathbf{l}\mathbf{r}')$ and $\mathbf{l}'^2$ into the metric (2.26), we arrive at the following form:

$$ds^2 = \mathbf{l}'^2(u)(v - \alpha(u)/\beta(u))^2 du^2 + ((\mathbf{l}\mathbf{r}')du + dv)^2,$$

from whence we already know how to pass to the standard form of the l. E. metric $dx^2 + dy^2$.

The necessity for the condition (2.27) is obtained using Theorem 3.2 presented further in Section 3.1. As is seen from this justification of the necessity, the proofs of the necessity for the previous two forms of l. E. metrics should not be easy.

Remark 2.6. The proof of the local Euclidity of the metric (2.26) under the condition (2.27) can be obtained by the direct search for the mapping of the form

$$x = A(u) + vB(u), \ y = C(u) + vD(u), \tag{2.28}$$

leading the metric (2.26) to the form $dx^2 + dy^2$. We obtain that the sought-for functions admit the following representations:

$$B(u) = \cos\gamma, D(u) = \sin\gamma, A' = \cos\varphi, \ C' = \sin\varphi$$

with the equalities

$$\cos(\varphi - \gamma) = (\mathbf{l}\mathbf{r}'), \ \gamma'^2(u) = \mathbf{l}'^2(u). \bullet \tag{2.29}$$

Open question 2.5. Let the l. E. metric ds^2 in the domain D of the variables (u, v) be defined as the metric of the surface S given by the explicit equation $z = z(U, V)$ and having a zero Gaussian curvature. If $S \in C^2$, then $ds^2 \in C^1$ and we know the formula for the isometric immersion of this metric into the standard Euclidean plane E^2. Can in this case a simpler formula be found and in which cases this metric admits an isometric immersion into E^2? In a different way, the question can be formulated as follows: is any explicitly given surface $z = z(x, y)$ with the flat metric an embedding of some domain of the standard Euclidean plane into E^3? $\bullet$

The obtained formulae are in practice inapplicable for studies of immersions and embeddings of domains with the metric (2.3) in the large into E^2, as it is difficult to formulate any simple condition to guarantee the single-valuedness of the mapping (2.4) in a multiply connected domain, and the more so its univalence even in a simply connected domain. Besides, the formulae (2.5)–(2.7) are inapplicable at a small regularity of the metric (2.3), say, for the case of $E, F, G \in C^{0,\alpha}, 0 < \alpha < 1$ (then the representation of (2.5) with (2.6) remains valid but the equations (2.7) for θ are missing). For this reason, we shall use another technique of reducing (2.3) to the standard form $dx^2 + dy^2$.

2.2.4 A characteristic feature of locally Euclidean metrics in weakly regular classes

As we have already said above, for the metrics (2.3) in any class of regularity there is an algorithmically admissible (via an infinite number of successive approximations, though) method of reducing them to the so-called isothermal or canonical form

$$ds^2 = \Lambda(\xi, \eta)(d\xi^2 + d\eta^2), \tag{2.30}$$

suitable for the domain D of any finite connection (see, for example, Chapter 2 in [184]). For this, we need to solve the Beltrami system

$$\frac{\partial \zeta}{\partial \bar{w}} = q(w)\frac{\partial \zeta}{\partial w}, \ \ \zeta = \xi + i\eta, \ w = u + iv, \tag{2.31}$$

where

$$q = (E - G + 2iF)/(E + G + 2\sqrt{\Delta}), \ \ \Delta = EG - F^2.$$

Suppose we found a solution $\zeta = \zeta(w)$ of the system (2.31), yielding the isometry of the domain D to some domain with the isothermal metric $\Lambda^2(\zeta)(d\xi^2 + d\eta^2)$, where

$$\Lambda^2(\zeta(w)) = \frac{\sqrt{\Delta}}{J(w)}, \Delta = EG - F^2, \tag{2.32}$$

and $J(w) = |\frac{\partial \zeta}{\partial w}|^2 - |\frac{\partial \zeta}{\partial \bar{\zeta}}|^2$ is the Jacobian of the mapping $\zeta(w)$. How can we know whether the obtained function $\ln \Lambda(\zeta)$ is harmonic or not? The thing is that although derivatives of $\Lambda(\xi,\eta)$ for ξ and η, probably, do exist, they cannot be directly computed in terms of E, F, G and $\zeta(w)$, because Λ, as is seen from (2.32), consists of a composition of functions having no second derivatives, if they were missing in the initial metric (2.3). For this reason, we shall first prove one criterion of harmonicity, requiring no calculation of the derivatives of the function.

Lemma 2.2. *For a real function $\lambda = \lambda(\zeta)$ of the class of smoothness $C^{0,\alpha}(\bar{\Omega})$, $0 < \alpha < 1$, given in the circle $\Omega : \xi^2 + \eta^2 \le R^2$, be harmonic, it is necessary and sufficient that it satisfies one of the following equations:*

$$\lambda(\zeta) = Re\left(\frac{R^2}{\pi} \iint\limits_{\Omega} \frac{\lambda(w)}{(R^2 - \zeta\bar{w})^2} du dv \right), \quad w = u + iv. \qquad (2.33)$$

$$\lambda(\zeta) = Re\left(\frac{1}{2\pi} \oint\limits_{\partial\Omega} \frac{\lambda(\tau)(\tau + \zeta)}{\tau(\tau - \zeta)} d\tau \right), \tau = Re^{i\psi}. \qquad (2.34)$$

Proof. The second criterion of harmonicity is evident if we remark that in the right-hand side of the equation (2.34) we have the Schwarz integral for expressing a holomorphic function in the circle through the values on the circumference of its real part [58]. Let us prove the first criterion.

Its sufficiency follows from the fact that in the right-hand side under the sign of Re we have a function, holomorphic in Ω, of the corresponding class of smoothness. To verify the necessity, we turn to the formula

$$T_\Omega(\zeta^n \bar{\zeta}^m) = -\frac{1}{\pi} \iint\limits_{\Omega} \frac{w^n \bar{w}^m}{w - \zeta} du dv =$$

$$\begin{cases} \dfrac{1}{m+1}\zeta^n\bar{\zeta}^{m+1} - \dfrac{1}{m+1}\zeta^{n-m-1}R^{2(m+1)}, & n \ge m+1 \\[2em] \dfrac{1}{m+1}\zeta^n\bar{\zeta}^{m+1}, & n < m+1, \text{ if } |\zeta| \le R \end{cases}$$

and

$$\begin{cases} 0, & n \geq m+1 \\[2em] \dfrac{1}{m+1}\zeta^{n-m-1}R^{2(m+1)}, & n < m+1, \text{ if } |\zeta| \geq R \end{cases}$$

(these formulae for the case of a unit circle are given on p. 44 in [184][7]).

Assume that $\lambda(\zeta)$ is a harmonic function. Then $\lambda(\zeta) = Re\Phi^+(\zeta) = \frac{1}{2}(\Phi^+ + \bar{\Phi}^+)$, where $\Phi^+(\zeta)$ is some function holomorphic in Ω. In the circle Ω we have

$$\Phi^+(\zeta) = \sum_{n=0}^{\infty} a_n \zeta^n.$$

As

$$\frac{R^2}{\pi}\iint\limits_{\Omega}\frac{w^n\bar{w}^m}{(R^2-\zeta\bar{w})^2}dudv = \frac{\partial}{\partial\zeta}\left(-\frac{1}{\pi}\iint\limits_{\Omega}\frac{w^n\bar{w}^m}{\bar{w}-\frac{R^2}{\zeta}}dudv\right),\quad |\zeta| < R,$$

the right-hand side of (2.33), by the formulae for $T_\Omega(\zeta^n\bar{\zeta}^m)$, is equal to

$$Re\left(\frac{R^2}{2\pi}\iint\limits_{\Omega}\frac{\Phi^+(w)+\bar{\Phi}^+(w)}{(R^2-\zeta\bar{w})^2}dudv\right) =$$

$$Re\left[\sum\frac{\partial}{\partial\zeta}\left(-\frac{1}{2\pi}\iint\limits_{\Omega}\frac{a_n w^n + \bar{a}_n\bar{w}^n}{\bar{w}-\frac{R^2}{\zeta}}dudv\right)\right] =$$

$$Re\left[\sum\frac{\partial}{\partial\zeta}\left(\frac{a_n}{n+1}(\frac{R^2}{\zeta})^{-n-1}R^{2(n+1)}\right)\right] = Re\Phi^+(\zeta) = \lambda(\xi,\eta).$$

The lemma is proved.

In the next lemma, we formulate a method of constructing the solution of the system (2.31), homeomorphically mapping the circle Ω onto itself.

---

[7] However, there is a typographic error there: the inequalities for n have $m-1$ everywhere instead of $m+1$; this error was kept in the 1988 edition.

Lemma 2.3. *Assume that the system (2.31) is given in the circle Ω : $|w| \leq R$ and the coefficient $q(w) \in C^{0,\alpha}(\bar{\Omega}), 0 < \alpha < 1$, and $q(0) = 0$. Then the unique solution $\zeta = \zeta(w)$ of the system, diffeomorphically mapping the circle $\bar{\Omega}$ onto the circle $\omega : |\zeta| \leq R$ with the conditions $\zeta(0) = 0, \dfrac{\partial \zeta(0)}{\partial w} > 0$, is presented as*

$$\zeta = we^{Q(w)}, \tag{2.35}$$

where $Q(w)$ is the solution of the equation

$$\frac{\partial Q}{\partial \bar{w}} = q(w)\frac{\partial Q}{\partial w} + \frac{q(w)}{w} \tag{2.36}$$

with the condition

$$ImQ(0) = 0, \ \ Re\,Q|_{\partial\Omega} = 0. \tag{2.37}$$

Finally, this solution Q can be found by successive approximations from formulae (2.38), (2.40), (2.42) given below.

Proof. In (2.36), the free term belongs to $L_p(\bar{\Omega}), |p - 2| < \varepsilon$ at a sufficiently small $\varepsilon > 0$. We shall search for a solution of the system (2.36) in the following form (cf. p. 210 in [184]):

$$Q(w) = -\frac{1}{\pi} \iint\limits_{\Omega} \left(\frac{h(\sigma)}{\sigma - w} + \frac{w\bar{h}(\sigma)}{R^2 - w\bar{\sigma}} \right) dadb + ic, \sigma = a + ib, Imc = 0. \tag{2.38}$$

The unknown function $h(w)$ will be sought for from the class $L_p(\bar{\Omega})$, $p > 1$. Then

$$\frac{\partial Q}{\partial \bar{w}} = h(w), \ \frac{\partial Q}{\partial w} = -\frac{1}{\pi} \iint\limits_{\Omega} \left(\frac{h(\sigma)}{(\sigma - w)^2} + \frac{R^2\bar{h}(\sigma)}{(R^2 - w\bar{\sigma})^2} \right) dadb \equiv \tilde{\Pi}\omega$$

should exist, and from (2.36) we have the equation for determining h:

$$h(w) = q(w)\tilde{\Pi}h + \frac{q(w)}{w}. \tag{2.39}$$

As the norm of the operator $\tilde{\Pi}$ in $L_2(\bar{\Omega})$ is equal to 1 (p. 218 in [184]) and $|q| \leq q_0 < 1$, then at some $p > 2$ the norm of $q(w)\tilde{\Pi}h$ in

L_p will be smaller than 1, so (2.39) admits the solution by successive approximations by the following scheme:

$$h_0 = \frac{q(w)}{w} \in L_p; \quad h_{n+1} = q(w)\tilde{\Pi}h_n + \frac{q(w)}{w} \in L_p, 2 < p < 2 + \varepsilon.$$

(2.40)

The limit function

$$h = \frac{q(w)}{w} + q(w)\tilde{\Pi}\left(\frac{q(w)}{w}\right) + q(w)\tilde{\Pi}\left(q\tilde{\Pi}(\frac{q(w)}{w})\right) + \ldots \quad (2.41)$$

belongs to L_p and determines by (2.38) a function $Q(w)$ with $Re\,Q(t) = 0, |t| = R$. In (2.38), we choose the term ic such that $Im Q(0)$ be equal to 0. We have

$$Q(0) = -\frac{1}{\pi} \iint\limits_{\Omega} \frac{h(\sigma)}{\sigma}\,dadb + ic,$$

whence

$$c = Im\frac{1}{\pi} \iint\limits_{\Omega} \frac{h(\sigma)}{\sigma}\,dadb.$$

The final form of $Q(w)$ is as follows:

$$Q(w) = -\frac{1}{2\pi} \iint\limits_{\Omega} \left(\frac{\sigma + w}{\sigma(\sigma - w)}h(\sigma) + \frac{R^2 + w\bar{\sigma}}{\bar{\sigma}(R^2 - w\bar{\sigma})}\bar{h}(\sigma)\right) dadb,$$

(2.42)

where h is given by the formula (2.41).

From (2.36) we have that $\zeta = we^{Q(w)}$ satisfies the system (2.31) (with the conditions given in the lemma), and because of this it belongs to $C^{1,\alpha}(\bar{\Omega})$. For $\zeta(w)$, it is also known that $\zeta(0) = 0$, $\dfrac{\partial\zeta(0)}{\partial\bar{w}} = \exp(Q(0)) > 0$; what is more, on the circle $|t| = R$ we have $|\zeta(t)| = R$. Let us show that $\zeta(w)$ is a homeomorphism of the circle $\Omega : |w| \le R$ onto the circle $\bar{\omega} : |\zeta| \le R$. Indeed, let $\zeta_0(w)$ be some solution of the system (2.31), homeomorphically mapping Ω onto itself with the condition $\zeta_0(0) = 0$ (such a solution always exists). Then any solution $\zeta(w)$ of the system (2.31) is representable as $\zeta = \Phi(\zeta_0(w))$, where Φ is a holomorphic function of ζ_0. In our case, the values of Φ for ζ_0 from the circle $|\zeta_0| = R$ also lie on the circle $|\zeta| = |\Phi| = R$ and $\Phi(0) = 0$. Because of this, Φ admits an analytical extension to the entire plane ζ_0, including the point $\zeta_0 = \infty$. However, this extension has only one zero, the point $\zeta_0 = 0$; and only one pole, the point $\zeta_0 = \infty$. Therefore, $\Phi = a\zeta_0$, so $\zeta(w) = a\zeta_0(w), |a| = 1$, which was required to be established. This reasoning also provides for the uniqueness of the mapping $\zeta(w)$ with the norming given in the lemma. The lemma is proved.

Now we can formulate the following criterion of the local Euclidity for the metrics (2.3), also suitable for metrics of class $C^{0,\alpha}, 0 < \alpha < 1$.

Theorem 2.3. *For the metric (2.3), given in the circle $\bar{\Omega} : u^2 + v^2 \le R^2$, be locally Euclidean, it is necessary and sufficient that the function*

$$\lambda(u, v) = \ln \Lambda = \ln(E + G + 2\sqrt{\Delta}) - \ln|e^{2Q}| - 2\ln\left|1 + w\frac{\partial Q}{\partial w}\right| - \ln 4$$

and the mapping $\zeta = we^{Q(w)}$ from Lemma 1.2 satisfy one of the following equalities:

$$\lambda(w) = Re\left(\frac{4R^2}{\pi} \iint\limits_{\Omega} \frac{\lambda(w_1)\sqrt{\Delta}|1 + w_1\partial Q/\partial w_1|^2|e^{2Q}|}{(R^2 - \zeta(w)\bar{\zeta}(w_1))^2} du_1 dv_1\right),$$

$$w_1 = u_1 + iv_1$$

$$\lambda(w) = Re\left(\frac{1}{\pi i} \oint\limits_{\partial\Omega} \frac{\lambda(t)(\zeta(t) + \zeta(w))}{\zeta(t) - \zeta(w)}\left(1 + \frac{q(t)R^2}{t^2}\right)\frac{\partial\zeta}{\partial t} dt\right),$$

$$t = Re^{i\psi} \in \partial\Omega.$$

Proof of the theorem is evident, if we recall that the local Euclidity of the metric $ds^2 = \Lambda^2(\zeta)(d\xi^2 + d\eta^2), \zeta = \xi + i\eta$, is equivalent to the harmonicity of the function $\lambda = \ln\Lambda$ with respect to the variables (ξ, η), and then use Lemma 2.3 and in (2.33) and (2.34) pass to the old variable w (herewith, for the Jacobian J we use the following expression: $J(w) = \dfrac{4\sqrt{\Delta}}{E + G + 2\sqrt{\Delta}}\left|\dfrac{\partial\zeta}{\partial w}\right|^2$).

Open question 2.6. For its verification, the presented criterion requires finding the limit of an infinite series of successive approximations of the functions $h_n(w)$. But these approximations are such that each successive approximation is "better" than the previous by its regularity properties, so there is hope that we can succeed in verifying the local Euclidity after *a finite number* of approximations, depending on the value of the Hölder index α in the condition for the metric (2.3).$\bullet$

2.3   *Isometric immersions and embeddings of locally Euclidean metrics into R^2. The case of a simply connected domain*

2.3.1   *Formula of isometric immersions for metrics in isothermal coordinates*

In works by many authors, for example, [184], it has been proven that the metric (2.3) can be reduced in the large to an isothermal form irrespective of whether it is given in a simply connected or multiply connected domain D. We suppose that the coefficients of the metric (2.3) are of class $C^{m,\alpha}(\bar{D}), m \geq 0, 0 \leq \alpha \leq 1, m+\alpha > 0$, and the boundary of the domain D is smooth of class $C^{m'+1,\alpha}, m' \geq 0, m'+\alpha > 0$. Then any locally diffeomorphic solution $\zeta = \zeta(w)$ of the system (2.31) possesses in the interior of the domain a smoothness of class $C^{m+1,\alpha}$ (for $\alpha \neq 0$ and 1, this is a corollary of the known theorems on the regularity of the solutions of an elliptical system, and for $\alpha = 0$ and 1 the proposed smoothness is a corollary of Theorem 1.1 on the smoothness of isometries from Chapter 1, because for l. E. metrics the coefficient $\Lambda(\xi, \eta)$ in (2.30) has an analytical smoothness). The situation with smoothness $\zeta(w)$ in the closed domain $\bar{D}$ is more complicated. The solutions $\zeta = \zeta(w)$ of the system (2.31), univalent or even single-valued in $\bar{D}$, may well have no solutions of class $C^{p+1,\alpha}(\bar{D}), p = min(m, m')$ (for $\alpha = 0$ or 1), but there always is a solution of class $C^{p+1,\alpha}(\bar{D})$ for $\alpha \neq 0$ or 1, and of class $C^{p+1,\beta}$ with any $\beta < 1$ for $\alpha = 1$, and of class $C^{p,\beta}$ with any $\beta < 1$ for $\alpha = 0$, diffeomorphically mapping D onto domain Ω of corresponding connection with the circular components of the boundaries. Further, the isometric mapping

$$x = x(\xi, \eta), \ y = y(\xi, \eta)$$

of the domain Ω with the l. E. metric (2.30) into the Euclidean plane E^2 with the coordinates (x, y) with account for the formulae (2.5)–(2.7) is determined from the relations $\varphi = 0$ and

$$x_\xi - y_\eta = 0, \ \ x_\eta + y_\xi = 0,$$

i.e., the isometric immersion of an l. E. metric in the isothermal form (2.30) into E^2 is given by a holomorphic function $z = \Phi(\zeta)$ with the condition

$$|\Phi'(\zeta)| = \Lambda(\zeta). \tag{2.43}$$

We prove the following sharpening of Theorem 2.1.

Theorem 2.4. *The simply connected domain D with an l. E. metric under conditions of smoothness $E, F, G \in C^{m,\alpha}(\bar{D})$ and $\partial D \in C^{m'+1,\alpha}, m \geq 0, 0 \leq \alpha \leq 1, m + \alpha > 0, m' \geq 0, m' + \alpha > 0,$ is isometrically immersed into E^2 by a mapping of smoothness $C^{m+1,\alpha}$ in D, which transfers ∂D into a curve of smoothness $C^{p+1,\alpha}$, $p = min(m, m')$.*

Proof. We need to prove this theorem, as a matter of fact, only for the cases of smoothness $E, F, G \in C^{0,\alpha}$ and $\partial D \in C^{1,\alpha}, 0 < \alpha < 1,$ because the other cases are already covered by Theorem 2.1 (we take into account that at $\alpha = 0$ we have $m \geq 1, m' \geq 1$).

We choose inside the domain D some point w_0 and, using, if necessary, the linear transformation of the variables, try to get in this point that $E(w_0) = G(w_0), F(w_0) = 0$, which will give us $q(w_0) = 0$ (the notation of the formula (2.31) is used). So as not to introduce new notation, we leave the old notation for the new variables and new domain. Now we map the domain D onto the circle $\tilde{D} : |\tilde{w}| \leq R$ by a conformal mapping $\tilde{w} = \tilde{w}(w)$ with the conditions $\tilde{w}(w_0) = 0, \tilde{w}'(w_0) > 0$. By Kellogg's theorem, this mapping shall be of class $C^{1,\alpha}(\bar{D})$, so the metric ds^2 transferred to the circle $\tilde{D}$ shall remain in class $C^{0,\alpha}$ in the closed circle, and for the new coefficients $\tilde{E}, \tilde{F}, \tilde{G}$ the old condition $\tilde{E}(0) = \tilde{G}(0), \tilde{F}(0) = 0$ will be satisfied. Now we can apply Lemma 2.3 and reduce the metric ds^2 from the circle $\tilde{D}$ to the isometric form (2.30), in which the variables ξ, η change in the circle $\bar{\Omega} : \xi^2 + \eta^2 \leq R^2$, and $\ln \Lambda(\xi, \eta)$ is a harmonic function of smoothness $C^{0,\alpha}(\bar{\Omega})$. In this class of smoothness, the conformal mapping $z = x + iy = \Phi(\zeta) : \bar{\Omega} \to E^2$ with the condition (2.43) is determined by the formulae

$$\Phi'(\zeta) = \exp\left(\frac{R^2}{\pi} \iint\limits_{\Omega} \frac{\ln \Lambda(w)}{(R^2 - \zeta\bar{w})^2} du dv\right) \in C^{0,\alpha}(\bar{\Omega}),$$

$$\Phi(\zeta) = -\frac{1}{\pi} \iint\limits_{\Omega} \frac{\Phi'(w)}{\bar{w} - \zeta} du dv \in C^{1,\alpha}(\bar{\Omega}). \qquad (2.44)$$

To prove the formula for $\Phi(\zeta)$, we use Lemma 2.2. We may suppose that $\Phi'(0) > 0$, as this can always be achieved by using, if necessary, some rotation of the plane (x, y). We know the real part of the holomorphic function $\ln \Phi'(\zeta)$, namely, $Re(\ln \Phi'(\zeta)) = \ln \Lambda(\zeta) \in C^{0,\alpha}(\bar{\Omega})$. Now using the first part of Lemma 2.2, we obtain

$$Re(\ln \Phi'(\zeta)) = Re\left[\frac{R^2}{\pi} \iint\limits_{\Omega} \frac{\ln \Lambda(w)}{(R^2 - \bar{w}\zeta)^2} du dv\right].$$

The integral in the right-hand side defines, according to [184], a function of the class $C^{0,\alpha}$; then, by the Privalov's theorem (see Chapter IX in [64]) we have $\Phi'(\zeta) \in C^{0,\alpha}(\bar{\Omega})$.

Now we prove the formula for $\Phi(\zeta)$. According to [184], the solution of the equation $\partial F/\partial\bar{\zeta} = f$ has the form

$$F(\zeta) = T_\Omega f + \frac{1}{2\pi i} \oint\limits_{\partial\Omega} \frac{F(t)}{t - \zeta} dt,$$

where

$$T_\Omega f = -\frac{1}{\pi} \iint\limits_{\Omega} \frac{f(u, v)}{w - \zeta} du dv.$$

We have $F(\zeta) = \bar{\Phi}(\zeta), f = \bar{\Phi}'(\zeta)$. Let us introduce the function

$$\Phi_1^-(\zeta) = \overline{\Phi(R^2/\bar{\zeta})}, |\zeta| \geq R, \Phi_1^-(\infty) = \overline{\Phi(0)}.$$

This function is holomorphic in the exterior of the circle Ω, and on its boundary $\Phi_1^-(t) = \overline{\Phi(t)} = F(t)$. Therefore, for $|\zeta| < R$ we have

$$\frac{1}{2\pi i} \oint\limits_{\partial\Omega} \frac{F(t)}{t - \zeta} dt = \frac{1}{2\pi i} \oint\limits_{\partial\Omega} \frac{\Phi_1^-(t)}{t - \zeta} dt = \overline{\Phi(0)},$$

i.e., accurate to a constant addend, $\Phi(\zeta)$ has the form proposed in (2.44). Following up the composition of the mappings $D \to \tilde{D} \to \Omega \to E^2$, we can see that it has the required class of regularity. The theorem is proved.

Remark 2.7. By analogy with Remark 2.5, we can assert that any open simply connected domain D with an l. E. metric (2.3) of smoothness $C^{0,\alpha}$, $0 < \alpha < 1$, admits an isometric immersion into E^2 of smoothness $C^{1,\alpha}$. Indeed, the conformal structure determined in D by the isothermal form of the metric (2.3) is conformally equivalent to the circle or the plane and, respectively, on D there is a mapping reducing the metric in the large to an isothermal form in the circle or on the plane. (This mapping will be holomorphic with respect to the conformal structure and of smoothness $C^{1,\alpha}$ with respect to the initial coordinates.) And when the l. E. metric in the large is written down in the isothermal form (2.30), its reduction to the standard form in the case of a simply connected domain is given in quadratures by curvilinear integrals.●

2.3.2   *Criteria of embeddability in the large*

Consider in more detail the problem of embeddings of the metric (2.30) into E^2, regarding (2.30) to be given in the circle $\bar{\Omega} : \xi^2 + \eta^2 \leq 1$. Generally speaking, there are corresponding criteria for the univalence of a function holomorphic in the circle, but they are rather cumbersome and geometrically low-informative, so now we will consider several sufficient conditions for the embeddability of the circle $\bar{\Omega}$ with the l. E. metric (2.30) into E^2.

a) Let the swerve of any arc of the boundary of the circle $\bar{\Omega}$ in the metric (2.30) be non-negative. Then $\bar{\Omega}$ embeds into E^2 as a convex domain.

This proposition can be considered as the corollary of the theorem on the cap-like realization of each convex domain on the manifold of non-negative curvature [1]; let us prove it by our method.

If the metric (2.30) is sufficiently regular, e.g., in the case of $\Lambda \in C^{1,\alpha}(\bar{\Omega})$, this proposition readily follows from the fact that the curvature k of the curve $\gamma : z = \Phi(t)$ is equal to the geodesic curvature

$$k_g = \frac{1}{\Lambda(t)}\left(1 + Re\frac{t\Phi''(t)}{\Phi'(t)}\right) = \frac{1}{\Lambda(t)}\left(1 + 2Re(t\frac{\partial \ln \Lambda}{\partial t})\right),$$

$$t = e^{i\psi} \in \partial\Omega, \quad (2.45)$$

of the edge $\partial\Omega$ of the circle $\bar{\Omega}$. As k_g is the ordinary curvature of the curve $\gamma : z = \Phi(t)$ and by the Gauss–Bonnet theorem

$$\oint k_g ds = 2\pi,$$

and $k_g \geq 0$ by the condition, then γ is a closed convex curve without self-intersections. Therefore, the mapping $z = \Phi(\zeta)$ is univalent at the boundary, so it diffeomorphically transfers the circle Ω into the interior of the curve γ.

b) On the boundary circumference, let $Re(t\Phi'(t)/\Phi(t)) \geq 0$. Then the circle $\bar{\Omega}$ with the metric (2.30), where $\Lambda \in C^{0,\alpha}(\bar{\Omega})$, can be embedded into E^2 in the form of a domain star-shaped relative to the point $z = \Phi(0) = 0$.

At $\Phi(\zeta) \neq 0$, $\zeta \neq 0$, the function $Re(\zeta\Phi'/\Phi) > 0$ harmonic in Ω is positive, and this is the necessary and sufficient condition for the function $\Phi(\zeta)$ regular in Ω to be univalent and to map Ω onto a domain star-shaped relative to the point $z = 0$ [64]. Let us show that this is also true without the *a priori* additional assumption that $\Phi(\zeta) \neq 0$ for $\zeta \neq 0$.

Consider the angle ψ between the tangent to the curve $\gamma = \Phi(\partial\Omega)$ and the position vector of the curve γ in the plane z. This angle is calculated by the simple formula

$$\psi = \arg t + \frac{\pi}{2} + \arg \Phi'(t) - \arg \Phi(t), \ t = e^{i\phi} \in \partial\Omega.$$

As the condition $Re(t\Phi'(t)/\Phi(t)) \geq 0$ implies that $\arg \Phi(t)$ increases at the positive-direction walking of point t along the circumference $\partial\Omega$, then $0 \leq \psi(t) \leq \pi$. Further, $\Phi'(t) \neq 0$, so $\Phi(t)$ may turn to zero at $\partial\Omega$ only in a finite number of points $t_1,\ldots,t_n$. At $t \to t_j^-$ the position vector $\Phi(t)$, being a secant, tends to the tangent; therefore, $\psi(t_j^-) = 0$. At $t \to t_{j+1}^+$ the tangent is directed away from $\Phi(t)$ to $\Phi(t_{j+1}) = 0$; and the secant, away from $\Phi(t_{j+1})$ to $\Phi(t)$, so $\psi(t_{j+1}^+) = \pi$. Let the position vector $\Phi(t)$ during the change of t from t_j to t_{j+1}

turn to a positive angle α_j, $1 \leq j \leq n$ $(t_{n+1} = t_1)$; then the tangent, as is seen from the above reasoning, shall turn to an angle $\alpha_j + \pi$; the general turn of the tangent shall be $\alpha_1 + \cdots + \alpha_n + n\pi$. But the general turn of the tangent is equal to the integral curvature of the curve γ, i.e., on the strength of (2.45), is equal to 2π. Therefore, either $n = 0$ and the turn of the tangent is equal to the turn of the position vector, which then should equal 2π, or else $n = 1$ and on γ there is one zero of the function $\Phi(t)$, and then the turn of the position vector equals π. In this case, γ is a curve without self-intersections, which passes through point $z = 0$. But $z = 0$ is also an image of the interior point $\zeta = 0$, whose neighbourhood is mapped into some neighbourhood of point $z = 0$, which is sure to go out of the domain inside γ. However, this is impossible as some boundary points of the image $\Phi(\Omega)$ should then be outside the interior of γ, but all of them coincide with the images of the boundary points.

Thus, on $\partial\Omega$ the function $\Phi(\zeta)$ does not go to zero. Because of this, γ is a smooth closed curve without self-intersections, containing the point $z = 0$ inside; herewith, the mapping $\Phi : \partial\Omega \to \gamma$ is bijective. And as the Jacobian of the mapping $z = \Phi(\zeta)$ is everywhere positive, it is easy to deduce from here that the mapping of the circle Ω onto the interior of γ is a homeomorphism.

This criterion of embeddability of the metric (2.30) is readily generalized for the case of the star property of γ relative to an arbitrary preset point $z_0 = \Phi(\zeta_0)$.

c) The condition from a) can be appreciably generalized, namely, it is sufficient to ask for the turn of any arc from $\partial\Omega$ to be greater than $(-\pi)$. In the case of $\Lambda \in C^{1,\alpha}$ this condition is tantamount to satisfying the inequality

$$\int_{\psi_1}^{\psi_2} k_g ds > -\pi$$

for any $0 \leq \psi_1 < \psi_2 \leq 2\pi$. The class of such curves was introduced in [96] under the name of almost convex curves, and [174] has proved that the smooth interior mapping (in the sense of Stoilov) of a circle with an almost convex boundary of an image is a homeomorphism ([19] gives another geometric description of this class of curves).

d) There are many conditions, to satisfy which is a necessary or sufficient criterion of univalence of the holomorphic functions in a circle, see [10, 64]. Each of these conditions shall also be an analytical criterion of embeddability of the metric (2.30) into E^2. As an example, consider the Nehari conditions (p. 547 in [64]). These conditions maintain that if a function $\Phi(\zeta)$ holomorphic in a unit circle is normalized by the equalities $\Phi(0) = 0$, $\Phi'(0) = 1$, then of two inequalities

$$|\{\Phi, \zeta\}| \leq \frac{6}{(1 - |\zeta|^2)^2}, |\{\Phi, \zeta\}| \leq \frac{2}{(1 - |\zeta|^2)^2}$$

the first is necessary and the second sufficient for the univalence of $\Phi(\zeta)$. Here $\{\Phi, \zeta\}$ denotes the Schwarz derivative for $\Phi(\zeta)$, which is equal to $\dfrac{\Phi'''}{\Phi'} - \dfrac{3}{2}\left(\dfrac{\Phi''}{\Phi'}\right)^2$. In our case,

$$\frac{\Phi''(\zeta)}{\Phi'(\zeta)} = 2\frac{\partial \ln \Lambda}{\partial \zeta},$$

so the Nehari conditions have, respectively, the form

$$\left|\frac{\partial^2 \ln \Lambda}{\partial \zeta^2} - \left(\frac{\partial \ln \Lambda}{\partial \zeta}\right)^2\right| \leq \frac{3}{(1 - |\zeta|^2)^2}$$

or

$$\left|\frac{\partial^2 \ln \Lambda}{\partial \zeta^2} - \left(\frac{\partial \ln \Lambda}{\partial \zeta}\right)^2\right| \leq \frac{1}{(1 - |\zeta|^2)^2}.$$

For functions univalent in a circle, there are also many estimates of their various characteristics (e.g., various estimates of Taylor coefficients, estimates of the function modulus increase, of the modulus of the derivative and its argument, the form of the range of values of some functionals related to $\Phi(\zeta)$ and $\Phi'(\zeta)$ etc. are known). Each of these estimates is a necessary condition of the embeddability of the metric (2.30) into E^2.

Let us illustrate the application of these conditions by an *example*. Consider in the circle $|\zeta| \leq 1$ the metric

$$ds^2 = \exp(2Re(a\zeta^n))(d\xi^2 + d\eta^2), \quad a = const, \quad n = 1, 2, \ldots . \quad (2.46)$$

For this metric

$$\Phi'(\zeta) = e^{a\zeta^n} = 1 + a\zeta^n + \frac{a^2\zeta^{2n}}{2!} + \dots$$

$$\Phi(\zeta) = \zeta + \frac{a\zeta^{n+1}}{n+1} + \frac{a^2\zeta^{2n+1}}{2!(2n+1)} + \dots .$$

The calculations yield

$$|\{\Phi, \zeta\}| = n|a||\zeta^{n-2}| \left| n - 1 - \frac{an}{2}\zeta^n \right|,$$

whence for $|a|$ as a necessary condition of embeddability of the metric (2.46) we obtain the following restriction:

$$|a| \leq$$

$$\inf_{0 \leq \rho < 1} \frac{12\rho^{2-n}}{n(1 - \rho^2)\left(\sqrt{(n-1)^2(1-\rho^2)^2 + 12\rho^2} + (n-1)(1-\rho^2)\right)},$$

$$\rho = |\zeta|^2.$$

Similarly, the sufficient Nehari condition yields that at

$$|a| \leq \inf_{0 \leq \rho < 1} \frac{4\rho^{2-n}}{n(1 - \rho^2)\left(\sqrt{(n-1)^2(1-\rho^2)^2 + 4\rho^2} + (n-1)(1-\rho^2)\right)}$$

the metric (2.46) is Euclidean in the large, i.e., admits an isometric embedding into E^2.

Let us calculate these restrictions more exactly. We obtain that at $n = 1$ for the Euclidity in the large of the metric (2.46) it is sufficient to have $|a| \leq 2$, and necessary to have $|a| \leq \sqrt{12}$ (the exact criterion: $|a| < \pi$), and at $n = 2$ we have as the sufficient condition the inequality $|a| \leq 1$ and as the necessary condition, $|a| \leq 2.85$. At $n \geq 3$, exact estimates are already technically difficult; however, at $n \to \infty$ we can find the main term of the restriction to $|a|$. Namely, the sufficient condition at large n has the form $|a| \leq \frac{4}{b^2 - 1} \exp(\frac{b^2 - 1}{2b}) + O(\frac{1}{n}) \approx 1.73 + O(\frac{1}{n})$, where b is the root of the equation $b^4 - 4b^3 - 1 = 0$, $b > 1$, and the necessary condition $|a| \le \dfrac{6}{b^2 - 3} \exp(\dfrac{b^2 - 3}{2b}) + O(\dfrac{1}{n}) \approx 2.34 + O(\dfrac{1}{n})$, where b is the root of the equation $b^4 - 4b^3 - 9 = 0$, $b > \sqrt{3}$. This necessary condition is much better than that given by the estimate $|a|^k < k!(nk + 1^2)$, $n = 1, 2, \ldots$, obtained by the proved Bieberbach hypothesis for coefficients of univalent functions.

The other properties of univalent functions can yield for $|a|$ more exact estimates, in which necessary and sufficient conditions are already rather close one to another. In particular, for necessary conditions, use can be made of the known estimates for the coefficients of univalent functions. For instance, with consideration of the restriction $|\Phi(\zeta)|$ by $(e^{|a|} - 1)/|a|$ and using the relation between the coefficients of univalent functions from [14], for $n = 2$ the necessary condition was successfully reduced to $|a| < \sqrt{7.5} \approx 2.74$; and at $n \to \infty$, to $|a| \le 2$. There are also other means to obtain either sufficient or necessary conditions, for example, using the condition of the almost convexity etc.

2.4 Isometric immersions of locally Euclidean metrics in the case of multiply connected domains

2.4.1 Formulae for isometric mappings

Let now the domain D be $(n + 1)$-connected. It can be considered that in passing from (2.3) to (2.30) $\bar{D}$ is mapped into a domain $\bar{\Omega}$ bounded by the circumference $\Gamma_0 : |\zeta| = R_0$, which contains pairwise exterior circumferences $\Gamma_j : |\zeta - \zeta_j| = R_j$, $1 \le j \le n$. Without loss of generality, we assume that $\zeta_1 = 0$. The orientation of the boundary $\partial\Omega$ is supposed to be positive.

Let us find the general form of the mapping $z = \Phi(\zeta)$, leading (2.30) to the standard form $dx^2 + dy^2$. The basis for this is the following:

Lemma 2.4. *Let the holomorphic function* $z = f(\zeta)$ *have in the domain* $\bar{\Omega}$ *a single-valued derivative* $f'(\zeta) \in C^{0,\alpha}$, *and let*

$$f'(\zeta) = \sum_{-\infty}^{+\infty} c_m^{(j)}(\zeta - \zeta_j)^m, \quad c_m^{(j)} = \frac{1}{2\pi i} \oint_{\Gamma_j^+} \frac{f'(\zeta)}{(\zeta - \zeta_j)^{m+1}} \, d\zeta, \quad 1 \le j \le n,$$

$$(2.47)$$

be expansions of the function $f'(\zeta)$ into the Loran series in the ring semi-neighbourhood of each circumference Γ_j, $1 \le j \le n$ (Γ_j^+ means that the contour Γ_j is passed around anticlockwise). Let

$$H(\zeta) = f'(\zeta) - \sum_{j=1}^{n} \frac{c_{-1}^{(j)}}{(\zeta - \zeta_j)}.$$

Then the function $f(\zeta)$ can be found by the formula:

$$f(\zeta) = \overline{\Phi_0(\zeta)} - \frac{1}{\pi} \int\int_{\Omega} \frac{H(w)}{\bar{w} - \bar{\zeta}} \, du dv + \sum_{j=1}^{n} c_{-1}^{(j)} \ln(\zeta - \zeta_j) + const, \quad (2.48)$$

where the function $\Phi_0(\zeta)$, holomorphic and single-valued in Ω, is given by the formula

$$\Phi_0(\zeta) = -\sum_{m=2}^{+\infty} \frac{\bar{c}_{-m}^{(0)} \zeta^{m-1}}{(m-1)R_0^{2m-2}} - \sum_{j=2}^{n} \bar{c}_{-1}^{(j)} \ln(R_0^2 - \zeta \bar{\zeta}_j) -$$

$$\sum_{j=1}^{n} \sum_{m=0}^{+\infty} \frac{R_j^{2m+2} \bar{c}_m^{(j)}}{(m+1)(\zeta - \zeta_j)^{m+1}} -$$

$$\sum_{j=1}^{n} \sum_{k \ne j} \bar{c}_{-1}^{(k)} \ln \frac{\zeta - \zeta_j}{(\zeta - \zeta_j)(\bar{\zeta}_k - \bar{\zeta}_j) - R_j^2}. \quad (2.49)$$

Proof. As

$$f'(\zeta) = H(\zeta) + \sum_{j=1}^{n} \frac{c_{-1}^{(j)}}{\zeta - \zeta_j}, \quad (2.50)$$

it is sufficient for us to find the holomorphic function $f_0(\zeta)$ with the derivative $f_0'(\zeta) = H(\zeta)$. By the definition of $H(\zeta)$, the integral of the single-valued function $H(\zeta)$ along any circumference $\Gamma_j, 0 \le j \le n$, is equal to zero; thus, $f_0(\zeta)$ shall also be a single-valued function. By the condition, we have

$$\frac{\partial \bar{f}_0}{\partial \bar{\zeta}} = \overline{H(\zeta)},$$

so by the formula for the general solution of an inhomogeneous Cauchy–Riemann system (p. 41 in [184]) we have:

$$\bar{f}_0(\zeta) = -\frac{1}{\pi} \iint\limits_{\Omega} \frac{\overline{H(w)}}{w - \zeta} \, du dv + \Phi_0(\zeta),$$

$$\Phi_0(\zeta) = \frac{1}{2\pi i} \oint\limits_{\Gamma} \frac{\bar{f}_0(t)}{t - \zeta} \, dt. \tag{2.51}$$

We show that in reality $\Phi_0(\zeta)$ can be found by using only the known function $H(\zeta)$.

From (2.51) and from the holomorphism conditions of $f_0(\zeta)$ and $H(\zeta)$, we have (p. 78 in [184]):

$$\Phi_0'(\zeta) = \frac{1}{\pi} \iint\limits_{\Omega} \frac{\overline{H(w)}}{(w - \zeta)^2} \, du dv = \frac{1}{2\pi i} \oint\limits_{\Gamma} \frac{\overline{H^+(t)}}{t - \zeta} \, d\bar{t}. \tag{2.52}$$

Now we remark that, to obtain the expansion of the function $f'(\zeta)$ into the Loran series in the neighbourhood of the circumference Γ_0, it is sufficient to know its expansions in the neighbourhood of each circumference Γ_j, $1 \le j \le n$. Indeed, in the neighbourhood of the circumference $\Gamma_0 : |\zeta| = R_0$ let us have

$$f'(\zeta) = \sum_{m=-\infty}^{+\infty} c_m^{(0)} \zeta^m, \quad c_m^{(0)} = \frac{1}{2\pi i} \oint\limits_{\Gamma_0^+} \frac{f'(\zeta)}{\zeta^{m+1}} \, d\zeta.$$

But the function $f'(\zeta)/\zeta^{m+1}$ is holomorphic and single-valued in Ω, so

$$\frac{1}{2\pi i} \oint\limits_{\Gamma_0^+} \frac{f'(\zeta)}{\zeta^{m+1}} \, d\zeta = \sum_{j=1}^{n} \frac{1}{2\pi i} \oint\limits_{\Gamma_j^+} \frac{f'(\zeta)}{\zeta^{m+1}} \, d\zeta.$$

We substitute here the values $f'(\zeta)$ from (2.47) on each circumference Γ_j^+ and by direct calculation find $c_m^{(0)}$:

$$c_{-1}^{(0)} = \sum_{j=1}^{n} c_{-1}^{(j)}$$

$$c_{m}^{(0)} = c_{m}^{(1)} + \sum_{j=2}^{n}\sum_{k=m}^{\infty}(-1)^{k-m}c_{k}^{(j)}C_{k}^{m}\zeta_{j}^{k-m}, \quad m > -1$$

$$c_{m}^{(0)} = c_{m}^{(1)} + \sum_{j=2}^{n}\sum_{k=1}^{-m}c_{-k}^{(j)}C_{-m-1}^{k-1}\zeta_{j}^{-m-k}, \quad m < -1,$$

where C_p^q as usually denotes the binomial coefficients.

By the formula (2.50) we now know the form of the function $H(t)$ on each circumference Γ_j, $0 \le j \le n$, and, thus, from (2.52) can find $\Phi_0'(\zeta)$ and then $\Phi_0(\zeta)$, and obtain the proposed formula (2.49). The logarithms involved in this formula are functions single-valued in Ω, as it is easy to verify that in Ω the arguments of the functions under the logarithms are single-valued. The lemma is proved.

It follows from the lemma that the function $f(\zeta)$ reconstructed in the multi-connected domain by its single-valued derivative shall be *single-valued in that and only that case, when all coefficients $c_{-1}^{(j)}, 1 \le j \le n$, are equal to zero.*

Let us now apply this lemma to describe the mapping $z = \Phi(\zeta)$, reducing the metric (2.30) to the standard form $dx^2 + dy^2$. In this case,

$$f'(\zeta) = (\ln \Phi'(\zeta))' = 2\frac{\partial \ln \Lambda}{\partial \zeta},$$

and, using the lemma, we find that

$$\Phi'(\zeta) = b\prod_{j=1}^{n}(\zeta - \zeta_j)^{c_{-1}^{(j)}}$$

$$\exp\left(\overline{\Phi_0(\zeta)} - \frac{1}{\pi}\iint\limits_{\Omega}\frac{2\frac{\partial \ln \Lambda(w)}{\partial w} - \sum_{j=1}^{n}\frac{c_{-1}^{(j)}}{w - \zeta_j}}{\bar{w} - \bar{\zeta}}\,du dv\right), \qquad (2.53)$$

where $\Phi_0(\zeta)$ is a function determined by the formula of the form of (2.49), and b is a constant determined to within the argument from the condition $|\Phi'(\zeta)| = \Lambda(\zeta)$ in a concrete point $\zeta \in \Omega$.

2.4.2   Isometric immersions

Now let us clarify when the mapping $\Phi(\zeta)$ with the derivative (2.53) shall yield an isometric immersion of the corresponding l. E. metric into E^2. For this, we first show that in (2.53) each coefficient $c_{-1}^{(j)}$ has a simple geometric meaning. Namely, as the tangent to the curve $z = \Phi(t)$ is inclined at an angle equal to $\varphi + \arg \Phi'$, $\varphi = \arg t$, $t \in \partial\Omega$, to the axis Ox, then

$$c_{-1}^{(j)} = \frac{1}{\pi i} \oint_{\Gamma_j^+} \frac{\partial \ln \Lambda}{\partial t}\, dt = \frac{1}{2\pi i} \oint_{\Gamma_j^+} (d \ln |\Phi| + id \arg \Phi') =$$

$$-1 + \frac{1}{2\pi i} \oint_{\Gamma_j} k_g\, ds.$$

Therefore, if the integral geodesic curvature of each circumference Γ_j^+, $1 \le j \le n$, is multiple to 2π, then $\Phi'(\zeta)$, as is seen from (2.53), shall be a single-valued function, and in this case $\Phi(\zeta)$ can again be found using the lemma. Herewith, the single- or many-valuedness of the function $\Phi(\zeta)$ depends on whether the coefficients $d_{-1}^{(j)}$ in the Loran expansion of the function $\Phi'(\zeta)$ in the neighbourhood of the circumferences Γ_j are equal to zero or not (cf. the relation of the single-valuedness of the function $f(\zeta)$ from the lemma with the equality to zero of the coefficients $c_{-1}^{(j)}$ for $f'(\zeta)$ from (2.47)). But if at least one of the circumferences Γ_j, $1 \le j \le n$, has an integral geodesic curvature not multiple to 2π, then $\Phi'(\zeta)$ and, together with it, $\Phi(\zeta)$ too are multi-valued functions and the immersion of (2.30) into E^2 is impossible.

We summarise in the form of the following:

Theorem 2.5. *For the l. E. metric (2.30) given in the multi-connected ring domain $\bar\Omega$ with the boundary $\Gamma = \Gamma_0 + \Gamma_1 + \cdots + \Gamma_n$ to admit an isometric immersion into E^2, it is necessary and sufficient that: 1) the integral geodesic curvatures (turns) of the circumferences Γ_j, $1 \le j \le n$, be multiple to 2π and 2) the coefficients $d_{-1}^{(j)}$ at $(\zeta - \zeta_j)^{-1}$ in Loran expansions of the function $\Phi'(\zeta)$ from (2.53) in the neighbourhood of Γ_j be equal to zero. Then the immersion*

$z = \Phi(\zeta)$ *is unique and is determined from (2.53) by analogy with the formula (2.48) from Lemma 2.4.*

We note that the conditions of the theorem do not require that the integral curvature of the interior circumference Γ_0 itself be multiple to 2π, as in the fulfillment of the conditions of the theorem it shall automatically be multiple to 2π by the Gauss–Bonnet theorem.

The absence of immersion of the metric (2.30) into E^2 can, therefore, be due to two reasons: 1) one of the coefficients $c_{-1}^{(j)} \neq 2k\pi$; 2) one of the coefficients $d_{-1}^{(j)} \neq 0$. Both these possibilities are realizable and give grounds to *classify* l. E. metrics unimmersible into E^2 as metrics of *conical* or *cylindrical* type. Therefore, by intrinsic geometry, *no special type corresponding to the torse – a surface of tangents to a spatial curve – is singled out* in locally Euclidean metrics as one of the three types of developable surfaces.

For non-integer $c_{-1}^{(j)}$, we could have given an analytical expression for the increments of $\Phi(\zeta)$ when passing along each boundary circumference Γ_j, but, because of their awkwardness and geometrically low informative value, we omit those formulae. We confine ourselves to detailed formulae only for the case of a doubly connected domain. For a concentric ring Ω, we have:

$$\Phi'(\zeta) = b\zeta^{c-1} \exp\left(\frac{1}{\pi} \int\int\limits_{\Omega} \frac{c_{-1} - 2w\frac{\partial \ln \Lambda(w)}{\partial w}}{w(\bar{w} - \bar{\zeta})}\, du\,dv - \right.$$

$$\left. \sum_{m=2}^{\infty} \frac{c_{-m}\bar{\zeta}^{m-1}}{(m-1)R_0^{2m-2}} - \sum_{m=0}^{\infty} \frac{c_m R_1^{2+2m}}{(m+1)\bar{\zeta}^{m+1}} \right),$$

$$\Phi(\zeta) = \frac{1}{\pi} \int\int\limits_{\Omega} \frac{d_{-1} - w\Phi'(w)}{w(\bar{w} - \bar{\zeta})}\, du\,dv + \sum_{m=2}^{\infty} \frac{d_{-m}\bar{\zeta}^{m-1}}{(m-1)R_0^{2m-2}} - $$

$$- \sum_{m=0}^{\infty} \frac{d_m R_1^{2+2m}}{(m+1)\bar{\zeta}^{m+1}} + d_{-1}\ln\zeta + const,$$

where

$$c_m = \frac{1}{\pi i} \oint\limits_{\Gamma_1^+} \frac{\partial \ln \Lambda}{\partial \zeta}\zeta^{-m-1}\, d\zeta, \quad d_m = \frac{1}{2\pi i} \oint\limits_{\Gamma_1^+} \frac{\Phi'(\zeta)}{\zeta^{m+1}}\, d\zeta.$$

Let $1 + c_{-1} = \dfrac{1}{2\pi} \oint\limits_{\Gamma_1} k_g\, ds$ be an integer.

If $d_{-1} \neq 0$, we get a multi-valued function $\Phi(\zeta)$ of the type of the logarithm (this corresponds to the multi-valuedness of a cylindrical type). But if c_{-1} is a non-integer, then both $\Phi'(\zeta)$ and $\Phi(\zeta)$ shall be multi-valued; herewith, $\Phi(\zeta)$ has the multi-valuedness of the same type as $\Phi'(\zeta)$, i.e.,

$$\Phi(\zeta) = \zeta^{c-1}\varphi_0(\zeta),$$

where $\varphi_0(\zeta)$ is a single-valued function (a multi-valuedness of a conical type is obtained).

Consider one example. Let the l. E. metric

$$ds^2 = \rho^{2b}(d\xi^2 + d\eta^2), \ \ \rho^2 = \xi^2 + \eta^2, \ \ b \text{ is a real number}, \qquad (2.54)$$

be given in the ring $\Omega : R_1 \leq |\zeta| \leq R_0$.

For this metric

$$\Phi'(\zeta) = \zeta^b, \ \ \Phi(\zeta) = \frac{\zeta^{b+1}}{b+1}, \ \ b \neq -1; \ \ \Phi(\zeta) = \ln\zeta, \ \ b = -1.$$

Then at an integer $b \neq -1$ the metric (2.54) can be immersed into E^2 in the form of a $|b+1|$ times covered ring, and at $b = -1$ there is no immersion, as $d^{(1)}_{-1} = 1 \neq 0$ (a univalent image of Ω is a rectangle with two identified opposite sides, which corresponds to a cylinder in E^3). At a non-integer b, there is no immersion either: a univalent image of Ω is a concentric part of the circular sector with a $2\pi|b+1|$ opening of the angle at the vertex, which corresponds to a cone in E^3.

In a multiply connected case, the l. E. metric unimmersible into E^2 can have a cylindrical type of unimmersibility in the neighbourhood of one or several circumferences; and on other circumferences, a conical type. Therewith, it turns out that in the intrinsic geometry of l. E. metrics there is *no* characteristic peculiar of torsial surfaces. Then it would be natural to ask the following:

Open question 2.7. Are there intrinsic features of the metrics given in multi-connected domains and immersible into R^3 in the large *only* as torses and unimmersible in the form of cylindrical or conical surfaces?•

Remark 2.8. The problem of finding a mapping $z = \Phi(\zeta)$ by $|\Phi'(\zeta)| = \Lambda(\zeta)$ and in simply and multiply connected cases is similar to inverse boundary problems in the theory of functions [58], so to check the univalence of $\Phi(\zeta)$ in a simply connected case, we can use the criteria of univalence established there. Note that the main distinction of the considered problem from inverse boundary problems is that in our case $|\Phi'(\zeta)|$ is known not only on the boundary but everywhere in the domain, so to find $\Phi'(\zeta)$ there is no need to search for the Green's functions of the domain Ω, whose finding in turn is an independent problem.●

2.4.3   Isometric embeddings

In the case of a multiply connected domain, the isometric embeddings of the metric (2.30) into E^2 are described by the following:

Theorem 2.6. *For the l. E. metric (2.30) of smoothness $C^{1,\alpha}$, $\alpha > 0$, given in a multiply connected circular domain Ω, to admit an isometric embedding into E^2, it is necessary and sufficient that the immersion $z = \Phi(\zeta)$ be univalent on each of the circumferences $\Gamma_j, 0 \leq j \leq n$.*

Indeed, the necessity of the condition is evident, and the sufficiency follows from the fact that under conditions of the theorem the l. E. metric (2.30) can be C^2-smoothly continued inside each circumference $\Gamma_j, 1 \leq j \leq n$ (provided that the integral geodesic curvature of each circumference Γ_j^+ is equal to 2π), which is easy to achieve, considering the Gauss–Bonnet formula and substituting, if necessary, the exterior circumference by a mapping of the form $z = \frac{1}{\zeta - \zeta_j}$. Then the l. E. metric continued to the whole circle $|\zeta| \leq R_0$ admits an immersion into E^2 with a univalent image of the boundary. This means that the immersion is an embedding.

2.4.4   Immersions using natural equations of boundary lines. The criterion of Jordanness of a closed curve

The criteria of immersibility and embeddability of the metric (2.30) into E^2 can be also obtained using the following geometric technique. For each component of the boundary of a multiply connected domain

Ω, we know the geodesic curvature of the metric (2.30), which at an isometric immersion into E^2 passes into the usual curvature of the curve γ – of the image $\partial\Omega$. Consequently, at the boundary the mapping $z = x + iy$ has the form

$$x = x_0 + \int_0^s (\cos \int_0^\sigma k_g(s)\, ds)\, d\sigma, \quad y = y_0 + \int_0^s (\sin \int_0^\sigma k_g(s) ds) d\sigma. \quad (2.55)$$

Then for the immersibility of the metric (2.30) it is *necessary and sufficient that the equalities* $x(0) = x(L_j)$, $y(0) = y(L_j)$ (where L_j is the length of the circumference in the metric (2.30)) be on each boundary circumference; and for the *embeddability (2.30) into* E^2, *that each curve* $\gamma_j : x = x_j(s)$, $y = y_j(s)$ – the image Γ_j – be a Jordan curve. However, the computation of $x(s)$ and $y(s)$ by (2.55) with account for (2.45) is exactly reduced to the formulae (2.48) and (2.53), so the formula (2.55) gives no advantage in practice. Apparently, it would be reasonable to use both variants of the solution depending on the problem. As an example, we present a reasoning, which makes it possible to propose an approach to the (long ago) formulated and (as yet) unsolved problem of finding the conditions for the curvature $k(s)$ as a function of the arc of the curve (see, for example, [6]), upon fulfilment of which the curve proves without self-intersections.

Let $k = k(s)$ be the curvature of the curve Γ in the function of its arc $s, 0 \le s \le L$. If the curve Γ is closed and without self-intersections, it bounds some simply connected domain D, which maps univalently conformally onto the unit circle Ω with the boundary $\gamma : \xi^2 + \eta^2 = 1$. The inverse function $z = \Phi(\zeta), \zeta = \xi + i\eta$ determines in Ω the locally Euclidean metric (2.30) with the coefficient $\Lambda(\zeta) = |\Phi'(\zeta)|$ or $\ln \Lambda = Re \ln \Phi'(\zeta)$. If we find $\Phi'(\zeta)$, and then also the mapping $\Phi(\zeta)$, the condition of its univalence is a necessary and sufficient condition of Jordanness of the curve Γ with the given curvature $k = k(s)$.

The geodesic curvature k_g of the circumference γ in the metric (2.30) coincides with the usual curvature k of the curve Γ. Let $s = s(\varphi)$ determine the mapping of the circumference γ on Γ, where φ is the polar angle, $0 \le \varphi \le 2\pi$. Then from the formula (2.45) we have that $\Phi'(t), t = e^{i\varphi}$, satisfies the following boundary condition:

$$Re[t\Psi'(t)] = \Lambda(t)k_g - 1, t = e^{i\varphi},$$

where the (holomorphic in the circle Ω) function $\Psi(\zeta) = \ln \Phi'(\zeta)$, and $k_g = k[s(\varphi)]$ with $ds = \Lambda d\varphi$. The solution of this boundary problem is given by the Schwarz integral, see [58]:

$$\zeta\Psi'(\zeta) = \frac{1}{\pi i} \oint_\gamma \frac{\Lambda k_g - 1}{\tau - \zeta} d\tau \qquad (2.56)$$

(the constant term $-\dfrac{1}{2\pi} \displaystyle\int_0^{2\pi} (\Lambda k_g - 1)d\varphi$ present in the general formula

is in our case equal to zero on the strength of the necessary condition

of Jordanness $\Gamma : \displaystyle\int_0^{2\pi} \Lambda k_g d\phi = \int_0^L k_g ds = 2\pi$). As

$$\oint_\gamma \frac{\Lambda k_g - 1}{\tau} d\tau = 0,$$

then, subtracting this integral from the right-hand side of the formula (2.56), we obtain that

$$\Psi'(\zeta) = \frac{1}{\pi i} \oint_\gamma \frac{\Lambda k_g - 1}{\tau(\tau - \zeta)} d\tau,$$

whence

$$\Psi(\zeta) = \ln \Phi'(\zeta) = \frac{1}{\pi i} \oint_\gamma \frac{\Lambda k_g - 1}{\tau} \ln \frac{\tau}{\tau - \zeta} d\tau + C. \qquad (2.57)$$

It is easy to verify that

$$\oint_\gamma \frac{1}{\tau} \ln \frac{\tau}{\tau - \zeta} d\tau = 0,$$

so from (2.57) with account for the equality $\Lambda = |\Phi'(\zeta)|$ we have

$$\Lambda(\zeta) = C \exp\left(Re\frac{1}{\pi i} \oint_\gamma \frac{\Lambda k_g}{\tau} \ln \frac{\tau}{\tau - \zeta} d\tau\right), \quad C = \Lambda(0) > 0. \qquad (2.58)$$

The equation (2.58) yields the representation of the coefficient $\Lambda(\zeta)$ of the locally Euclidean metric (2.30) in a unit circle through the values of Λ at the boundary and the geodesic curvature of the boundary. The problem of using this formula in our case is that we do not know the function $s = s(\varphi)$, which is an argument for $k_g = k[s(\varphi)]$. Let us introduce the inverse function $\varphi = \varphi(s)$ with $\varphi(0) = 0, \varphi(L) = 2\pi$. As $\Lambda(\tau(s))d\varphi = ds$ and $d\tau/\tau = id\varphi$, then, passing in (2.58) to the limit at $\zeta \to t \in \gamma$, we obtain the equation for $\Lambda(t)$:

$$\Lambda(t) = C \exp\left(Re\frac{1}{\pi} \int_0^L k(\sigma) \ln \frac{e^{i\varphi(\sigma)}}{e^{i\varphi(\sigma)} - t} d\sigma\right), t = e^{i\varphi(s)}, \tau = e^{i\varphi(\sigma)}.$$

$$(2.59)$$

As $\Lambda d\varphi = ds = s'(\varphi)d\varphi = d\varphi/\varphi'(s)$, then $\Lambda(s) = 1/\varphi'(s)$; hence, we have

$$\varphi'(s) = \frac{1}{C} \exp\left[Re\left(\frac{-1}{\pi} \int_0^L k(\sigma) \ln \frac{\tau}{\tau - t} d\sigma\right)\right]. \qquad (2.60)$$

We reassign t in (2.59) as $t_1 = e^{i\psi(s_1)}$ and, multiplying both parts of the equality by $d\psi$ and finding the constant C from the condition $\int_0^{2\pi} \Lambda(t_1)d\psi = L$, substitute the found value of C into (2.60). The following relation is obtained:

$$\varphi'(s) = \frac{1}{L} \int_0^{2\pi} \exp\left(Re\frac{1}{\pi} \int_0^L k(\sigma) \ln \frac{\tau}{\tau - t_1} d\sigma\right)d\psi$$

$$\exp\left[Re\left(\frac{-1}{\pi} \int_0^L k(\sigma) \ln \frac{\tau}{\tau - t} d\sigma\right)\right].$$

As the second multiplier here does not depend on ψ, it can be inserted under the first integral sign, and we arrive at the following integro-differential equation for $\varphi(s)$:

$$\varphi'(s) = \frac{1}{L}\int\limits_0^{2\pi} e^{\mathrm{Re}\left(\frac{1}{\pi}\int\limits_0^{L} k(\sigma)\ln\frac{e^{i\varphi(\sigma)}-e^{i\varphi(s)}}{e^{i\phi(\sigma)}-e^{i\psi}}d\sigma\right)}\,d\psi \qquad (2.61)$$

with the boundary conditions $\varphi(0) = 0, \varphi(L) = 2\pi$. If we solve the equation (2.61) for $\varphi(s)$ with the given boundary conditions (when solving, we need to remember that the known function $k(s)$ necessarily satisfies the condition $\int\limits_0^L k(s)ds = 2\pi$), then, finding C from (2.60) and $\Lambda(t)$ from (2.59), we calculate from (2.58) $\Lambda(\zeta)$ and by the formula (2.44) find the sought-for mapping $\Phi(\zeta)$. *Its univalence is a necessary and sufficient condition for the Jordanness of the curve* Γ *with a given curvature* $k = k(s)$. Note that we simultaneously indicated the method of finding a conformal mapping of the circle onto a domain bounded by a Jordan curve Γ. But, certainly, studying the resolvability and finding the solution of the non-linear equation (2.61) is not an altogether easy task.

Solving the equation (2.61) followed by finding $\Lambda(\zeta)$ by the formulae (2.59) and (2.58) can be considered to be the solution of the problem of determining an l. E. metric in the circle by the geodesic curvature of the boundary of the circle, see [148].

2.5 Standard representations of two-dimensional spatial forms

2.5.1 Complete locally Euclidean metrics

The metrics of the five known two-dimensional forms – a plane, a cylinder, a torus, a Möbius band and a Klein bottle – can also be given not in the standard form. For instance, the metric of a plane can be given in a circle $D : u^2 + v^2 < 1$ in the form of (2.3) with the additional condition of its completeness. Then by general theory we can say that such a metric is reduced to the standard form $dx^2 + dy^2$ with the change of (x, y) over all the plane, but the explicit formulae of mapping the circle D onto the plane can be written via the curvilinear integrals using the relations (2.5)–(2.7) only in the case of C^1-smoothness of the metric given in D. However, before the search for the very isometry to the plane it is natural to put such an

Open question 2.8. How can it be determined by the coefficients of the metric (2.3) whether this metric is complete?•

Some answers to this question can be obtained from the results by V.A. Alexandrov presented in Chapter 6 (to use them, the metric should be transferred to the entire plane, e.g., by the mapping $x = \dfrac{u}{1-r^2}, y = \dfrac{v}{1-r^2}, r^2 = u^2 + v^2$). It can only be said at once that a metric complete in the circle cannot be given in isothermal coordinates, because otherwise the inverse mapping from the plane into the circle would have been given by a bounded holomorphic function on the plane. Evidently, it cannot be given in Chebyshev and polar geodesic coordinates, either.

Open question 2.9. Can a complete metric in the circle be given in orthogonal coordinates?•

In order to solve this problem, one can try to use the following approach: to write down the system (2.1) for the components of the inverse mapping $(x, y) \to (u, v)$ and to study the obtained elliptic system for the existence of a solution bounded on the entire plane. The existence of such a solution is impossible at a uniform ellipticity of the system, and an additional study is required under conditions of non-uniform ellipticity as in our case.

As for the reduction to the standard form of l. E. metrics corresponding to the other two-dimensional spatial forms and given by l. E. metrics in the form (2.3), here virtually all the range of problems can be represented only as:

Open question 2.10. Let l. E. metrics in the form of (2.3) be given in some domains. How can we describe the conditions under which these domains after respective identification of the boundaries' segments can be converted to domains with complete l. E. metrics of the topological structure of one of the four types Euclidean forms – a cylinder, a torus, a Klein bottle or a complete Möbius band; and if it has been done or if it is known beforehand, then how can an isometric image be constructed on respective canonical domains?•

For some cases of a cylinder-type metric, the answer can be obtained from the immersion of a cylinder into E^3, presented at the end of Subsection 4.3.3.

2.5.2 Extension of locally Euclidean metrics

There is one more problem of the theory of l. E. metrics, which seems to have been ignored in the literature altogether. This is the problem of extension of l. E. metrics. Let in some domain D with the boundary $\partial D \in C^{n,\alpha}, n \geq 1, 0 < \alpha < 1$, an l. E. metric (2.3) of regularity $C^{n-1,\alpha}(\bar{D})$ be given with the condition $EG - F^2 > 0$ in $\bar{D}$. Then locally in the neighbourhood of each point of the boundary it admits extension over the boundaries of the domain D with the preservation of the property of local Euclidity. Indeed, let M_0 be an arbitrary point of the boundary. We can map the neighbourhood of the point M_0 into the plane xy such that the metric take the standard form $dx^2 + dy^2$; herewith, the sufficiently small arc $\gamma \subset \partial D$ passes into some curve Γ on the plane xy of smoothness $C^{n,\alpha}$, and points of the neighbourhood M_0 inside the domain D pass into some domain T^-, say, lower than the curve Γ. The functions $u = u(x, y)$, $v = v(x, y)$ of smoothness $C^{n,\alpha}$, mapping the domain T^- into the considered neighbourhood of the point M_0, can be extended $C^{n,\alpha}$-smoothly to some domain T^+ higher than the curve Γ, and as in this case the Jacobian of the mapping $(x, y) \to (u, v)$ in the pre-image of the point M_0 is not equal to zero, the constructed mapping will be a diffeomorphism in some domain $T \subset T^- \cup T^+$ and will translate the Euclidean metric $dx^2 + dy^2$ into some complete neighbourhood of the point M_0 with the coincidence of the translated metric inside D with the metric ds^2 already available there. If the domain D admits an isometric immersion into the plane, then the described extension is possible along the entire curve ∂D; what is more, it is, evidently, not unique, because in the domain T^+ it is possible to take any l. E. metric, sufficiently smoothly extending the standard metric from the domain T^-. For this reason, in reality it is of interest to consider the following:

Open question 2.11. If some l. E. metric is given in the domain D, what is the maximum extension of the metric with preservation of local Euclidity?•

To illustrate the non-triviality of this problem, consider three examples.

1) Let the domain D with an l. E. metric admit an isometric embedding into the plane. Then its image on the plane can be extended to the whole plane and, correspondingly, D can be extended to a simply connected domain with a complete l. E. metric.

2) Let the simply connected domain D with an l. E. metric be isometrically immersible but not embeddable into the plane. Then the domain D could not be extended *a fortiori* to some simply connected domain D^* such that the metric in it becomes complete. Indeed, otherwise the domain D^* with an l. E. metric would have admitted an isometric embedding into the plane, and then the limitation of this embedding onto the domain $D \subset D^*$ would have been an isometric embedding of D into the plane, which is impossible by the condition. Therewith, in this case the description of a maximum possible extension of the l. E. metric into D requires to be additionally studied with consideration of particular properties of the given metric.

3) Consider the metric $ds^2 = r(du^2 + dv^2), r^2 = u^2 + v^2$, given in the domain $D : 0 < R^2 \leq u^2 + v^2 < \infty$. The integral geodesic curvature of the circumference $r = const$ is not equal to zero, and, therefore, this metric in no class of regularity can be extended inside the circle $u^2 + v^2 < R^2$ as an l. E. metric.

3   Surfaces with locally Euclidean metrics

3.1   *External structure in smoothness class C^1*

Surfaces with locally Euclidean metrics have different descriptions with respectively different names (a list with characteristic properties can be found, for example, in [182]), which in reality do not always reflect the same class of surfaces, as indicated in earlier times by Lebesgue ([104], pp. 328–330). Indeed, they are largely presented as ruled surfaces composed of four types of domains – a plane domain, a cylinder, a cone and a development of tangents to some spatial curves. However, there are surfaces with l. E. metrics, whose geometry does not involve the customary rectilinear generatrices. And inversely, there are ruled surfaces (even cylinders and cones), along the generatrices of which the tangent plane to the surface does exist and is constant, but the surface has no l. E. metric ([104], p. 333).

3.1.1   *Definition of surfaces with locally Euclidean metrics and the general case of C^1 smoothness*

The starting point of the general determination of surfaces with l. E. metrics should be manifolds M_0 with an intrinsic metric locally isometric to the metric of the Euclidean plane[8]. Then any mapping $f : M_0 \to E^3$ preserving the lengths of the curves (as always, distinguishing in $f(M_0) \subset E^3$ the spatially coinciding points with different pre-images and understanding the curves on the surface as images of the curves on the manifold M_0) can be called a surface with an l. E. metric. In particular, then any rectifiable curve in M_0 passes into a rectifiable curve of the same length on the surface and, respectively, a non-rectifiable curve passes into a non-rectifiable one. This enables the assertion that the mapping f, while remaining in respective sets, locally preserves the distances between points in M_0 and $f(M_0)$ (understood to be the lower bounds of the lengths of the curves connecting these points).

Remark 3.1. It is worth noting that in a general case the requirement of keeping the lengths of *only rectifiable* curves when mapping f may prove insufficient for obtaining an isometry, because then not every rectifiable curve in $f(M_0)$ should be a mapping of some rectifiable curve from M_0. The possibility of such a situation in principle is confirmed by the following example: let a cylinder $S_0 : \{x, y(x), 0\} + t\{0, 0, 1\}$, $a \leq x \leq b$, $-1 \leq t \leq 1$, over a curve $y = y(x)$ non-rectifiable on any part be mapped onto a cylinder $S : \{x, Y(x), 0\} + t\{0, 0, 1\}$, $a \leq x \leq b$, $-1 \leq t \leq 1$, with a rectifiable directrix $y = Y(x)$ with the generatrices changing into those by the equality of parameters x and t; then the only rectifiable lines on S_0 – segments on the generatrices – pass into rectifiable lines of the same length on

---

[8] Manifolds with an l. E. metric appear in the class of two-dimensional manifolds with an intrinsic metric of bounded curvature in the sense of Alexandrov (MBC), in this case with the value of the curvature function $K \equiv 0$, see [3]. As on two-dimensional MBC it is always possible to introduce isothermal coordinates [136], in which the l. E. metric has an analytical class of smoothness, we can consider that locally M_0 is a two-dimensional domain with the standard Euclidean metric.

S, and the rectifiable directrix on S is an image of the non-rectifiable directrix on S_0, whereas the mapping fails to yield an isometry of S_0 on S. Most probably, such a situation in the case of a mapping from the l. E. manifold M_0 is impossible, but "just in case" we required the mapping f to preserve both the finite and infinite lengths. ●

Remark 3.2. To determine the surface $f : M_0 \rightarrow E^3$ with an l. E. metric, we could have apparently first introduced a metric on $f(M_0)$ as a distance between points, determined by the lower bound of the lengths of the curves on the surface as curves in ambient space. Then we would have required a local isometricity of the metrics on M_0 and $f(M_0)$. But in that case, getting the property of preserving the lengths of rectifiable curves would have required tedious analysis of the relation between the definitions of the length of the curve on the surface through the lengths of inscribed *spatial* broken lines (which is required by the definition of the length of a curve in space) and linked curves composed by shortest arcs inscribed into its pre-image on the manifold M_0 (which corresponds to the definition of the length of a curve on M_0)[9]. For this reason, we preferred the definition of local Euclidity of a surface straight by the equality of the lengths of curves in M_0 and their images on $f(M_0)$. ●

If a surface is obtained as a mapping in E^3 of some plane domain D of the change of parameters, which transfers any rectifiable (in Euclidean sense) curve in D into some rectifiable curve in the image with the possible change of length of no more than some fixed number of times, then, following Lebesgue [104, 105], it is called a *rectifiable* surface. In [105] Lebesgue showed that rectifiable surfaces have a tangent plane almost everywhere. Consequently, for surfaces with an l. E. metric there is some regularity already in the most general case.

However, without any additional requirements such surfaces can be absolutely different from the usual form of "developable" surfaces – there may be not a single rectilinear segment on them. In [104], Lebesgue gave an example of such a surface. For this, we should take a usual truncated circular cone $C : X = x,\ Y = y, Z = \sqrt{x^2 + y^2}$

---

[9] In the case of an isometry between Riemannian spaces, such an analysis is not required, as in both spaces the lengths of the curves are defined by the upper bound of the sums of *intrinsic* distances by finite systems of points on the curves.

72   I.Kh. SABITOV

between the planes $z = z_0$ and $z = z_1$ and, on its basis, construct a surface $S : X = x, Y = y, Z = \Psi(\sqrt{x^2 + y^2})$, where $\Psi(z)$ is a non-linear continuous function of bounded variation, having at each segment $[z_0, z], z_0 < z \leq z_1$, the total variation equal to $|z - z_0|$ (such a function is constructed, for example, in [104], pp. 289–290). This function has a minimum and a maximum at each interval, and its graph contains no rectilinear segments whatsoever. The length of the curve $X = X(t), Y = Y(t), Z = Z(t)$ is determined through total variations of its components, so any rectifiable curve on C passes into a rectifiable curve of the same length on S; therefore, the surface S is locally isometric to a domain on the plane; however, remaining a surface of revolution, it shall contain *a fortiori* no segments of a straight line.

Henceforth, under a "developable" surface we will understand any continuous connected surface with a locally Euclidean metric.

With some rather general additional requirements imposed on a surface, the geometry of developable surfaces admits a rather detailed description. We shall start, however, with the assumption that the only thing known about the surface is that it is of smoothness C^1 (note that in the examples by Lebesgue the surfaces are not C^1-smooth). By the known Nash–Kuiper theorems ([112] and [103]), such isometric immersions always exist, and as these surfaces can be encased into a sphere of arbitrarily small radius by bending in the same class of surfaces, it can be asserted that they have no rectilinear generatrices preserved in the process of bending. Moreover, in [197] it has been proved that there exist C^1-smooth developable surfaces, where some shortest arcs have no half-tangents and enter the considered point in the form of a spiral; consequently, no rectilinear segment whatsoever passes through this point. Some sharpening of the class of regularity of the surface is of no help: in [28], it has been proved that, by bending the plane, it is possible to construct surfaces of class $C^{1,\alpha}, \alpha < 1/13$, on which there are no rectilinear generatrices (recently, these results have been presented in a more full and unified form in [29], which, referring to V.A. Zalgaller, specifies how the boundary of $\alpha < 1/13$ could be brought to the value of $\alpha < 1/7$).

3.1.2   The Pogorelov class of C^1-smooth surfaces

Now we assume that the surface $S = f(M_0)$, apart from the smoothness of class C^1, also has a bounded extrinsic curvature in the Pogorelov sense (BECP) (see the book [124]). (Demonstrably, this condition implies that the spherical image of the surface has a finite "geometric" (non-oriented) area, in calculation of which the areas of the regions with different orientations on the spherical image are not mutually cancelled but are added up.) Then by the Pogorelov theorem ([124], p. 694) we have that for each point $Q \in S$ there are two variants: either the point Q has on S a neighbourhood, which is a plane domain; or a *unique* rectilinear segment passes through Q, with endpoints at the boundary of the surface, along which the tangent plane to S is stationary. In the class of C^1-smooth BECP surfaces, the structure of surfaces with the complete l. E. metric is also known – all of them are cylindrical surfaces [124].

The points being interior points of plane domains on the surface shall be called *planar* points of the surface (not to be confused with points of *flattening*, which are determined by the condition of the equality of all coefficients of the second quadratic form in them to zero, and by the definition itself require to be assigned to a surface of class C^2 – any planar point is a point of flattening, but the inverse is not true[10]). Each connected component of a set of planar points is an open set. We call a *relative boundary* of some set A on an open surface its boundary points, not being the boundary points for S; speaking otherwise, the relative boundary of the set A consists of those of its boundary points, which are interior points of S. The point lying on the relative boundary of a plane domain is not planar, so the unique rectilinear generatrix passes through it. Then, the structure of plane domains on S is described as follows.

---

[10] English-speaking literature has no unified terms for such kind of points, cf. e.g., the terminology of [172], p. 628, [173], p. 427 and [77], p. 767; in Russian-speaking literature, there is a generally accepted concept of the "point of flattening"; for this reason, for a particular case when a given point of flattening has a neighbourhood consisting completely of points of flattening and, therefore, is an interior point of some plane domain, we chose the name "planar point", which, as we believe, by its primary meaning agrees well with the nature of such a point.

Property 3.1. *The relative boundary of any connected plane domain on an open developable surface S of the BECP class consists of a finite or countable number of rectilinear intervals running up to the boundary of the surface.*

As a consequence, we can assert that *the relative boundary of a plane domain cannot contain an interior point of the surface, in which two rectilinear parts of the boundary intersect*, i.e., each connected component of the relative boundary of the plane domain on S is a rectilinear interval, and the possible vertices of the boundary of the plane domain should be boundary points of the surface S itself.

3.1.3 C^1-smooth torses

By the terminology dating well back to Euler, a surface possessing the property that at least one rectilinear segment passes through each of its points[11], along which the tangent plane to the surface is the same, or, as the phrase goes, the tangent plane is stationary along the generatrix, is called a *torse* (apparently, there is no such clear-cut definition by Euler; we gave it following [77]). We cannot declare each torse to be a surface with an l. E. metric, as there exist, for example, cylinders with a non-rectifiable directrix located on the cross section orthogonal to the generatrices and having a tangent at each point (and, herewith, the tangent plane to the surface along each generatrix exists and is constant but not continuous on the surface). Also, for the cylinder to be locally isometric to the plane it is necessary and sufficient that its directrix orthogonal to the generatrices be rectifiable ([104], p. 333) (similar constructions are also possible for cones and surfaces formed by tangents to a spatial curve). To rule out this example, we will assume that *only C^1-smooth torses* are considered. If it is additionally assumed that we consider C^1-smooth surfaces of only BECP class, then *each torse has an l. E. metric and, moreover, torses in this BECP class represent a complete description of surfaces with l. E. metrics.* Indeed, by the above mentioned Pogorelov theorem, each C^1-smooth surface with bounded extrinsic curvature equal to zero is a torse. Conversely, if a C^1-smooth torse of BECP class is

---

[11] The expression "a segment passes through a point" means that the point considered is an *interior point* of this segment.

given, all of its points are irregular in the sense of [124], Chapter IX, so its extrinsic curvature in the Pogorelov sense is zero. In the same book ([124], Chapter IX), it has been proved that BECP surfaces are manifolds of bounded curvature in the Alexandrov sense and their extrinsic and intrinsic curvatures coincide, therefore, torses of BECP class have zero intrinsic curvature, i.e., an l. E. metric. Thus, we proved the following:

Theorem 3.1. *For a C^1-smooth surface of class BECP to have an l. E. metric, it is necessary and sufficient for it to be a C^1-smooth torse.*

As, starting from the "classical" smoothness C^2, surfaces fall under the conditions of the Pogorelov theory on BECP surfaces, from Theorem 3.1 we have:

Corollary 3.1. *For a surface of smoothness class C^2 to have an l. E. metric, it is necessary and sufficient for it to be a torse.*

In [77], the proposition of this corollary was proved with an additional condition that all points of the surface of a torse are of one type: either all of them are points of flattening or else it has not a single point of flattening. Without this condition, the corollary was proved in [76]; the same work noted it to follow also from the results by Pogorelov, which we have demonstrated.

Remark 3.3. In Theorem 3.1, the *a priori* requirement for the structure of torses can be weakened, keeping their property to have an l. E. metric. Namely, in the definition of torses instead of the condition that "at least one rectilinear segment, along which the tangent plane is 'stationary', *passes* through each point" it could only be assumed that "at least one rectilinear segment with the stationary tangent plane *emanates* from each point". Then, as before, there will be no regular points (by the terminology of the book [124], Chapter IX) on such surfaces of BECP class, so they will be surfaces of extrinsic zero curvature in the Pogorelov sense, which means of zero intrinsic curvature. Furthermore, by the cited Pogorelov theorem the unique generatrix running from edge to edge of the surface will pass through each nonplanar point; consequently, torses with "weakened" conditions would be, in reality, of previous structure (provided that they are of class BECP).•

3.1.4 C^1-smooth torses and normal developable surfaces

The previous theorem is insufficiently good in that torses are requested *a priori* to be of the class of BECP surfaces. In the mean time, it turns out that all C^1-smooth torses have an l. E. metric, and, what is more, there is a class of C^1-smooth surfaces, to which all C^1-smooth torses belong and in which all surfaces with an l. E. metric are torses. We begin by presenting the definition of torses in another way. For this, we introduce, following [159], see also [32], p. 82, the class of normal developable surfaces in E^3. Their definition is given in two steps. First, we define normal surfaces: a C^1-smooth surface S is called a *normal* surface of non-negative curvature, if for each point M one of the following conditions is satisfied: (a) the point M has a neighbourhood in the form of a convex surface (in particular, this neighbourhood can be a plane domain); (b) a rectilinear generatrix passes *through* the point M, and along the generatrix the tangent plane to the surface is stationary. Then a *normal developable* surface is such a normal surface of non-negative curvature, which has no points of strict convexity. Therefore, for any point on a normal developable surface there are two variants: either a rectilinear generatrix with the stationary tangent plane passes through it, or the point has a neighbourhood in the form of a nonstrictly convex surface. In the second case, the surface in the neighbourhood of the considered point is of BECP class and then its extrinsic curvature in this neighbourhood should be zero, because no domain with positive extrinsic curvature can be realized in the form of a nonstrictly convex surface. And we already know that on any surface of BECP class with zero curvature each point is either planar, or else a generatrix with the stationary tangent plane passes through it. In the case of a plane neighbourhood, a rectilinear segment along which the tangent plane is constant also passes through the point (such segments are in reality infinitely many). Thus, *any normal developable surface turns out to be a torse.*

Conversely, let a C^1-smooth torse S be given; let us show it to be a normal developable surface. For this, it is sufficient to make certain that there are no points of strict convexity on it. Indeed, assume the opposite, and let M be a point of strict convexity. By the definition of the point of strict convexity, there exists a plane locally supporting to S at point M and intersecting in its sufficiently small neighbourhood with S only at point M. Consequently, no segment of a straight line can pass through point M – this is a contradiction.

Let us now study some properties of torses, which are well known for the case of surfaces of class C^2.

Property 3.2. If some point A on a torse S is not planar, then any other point on the generatrix passing through this point with the same tangent plane preserved along this generatrix is not planar either.

Let there be a planar point B on the generatrix l_A passing through some non-planar point A. First we will consider the case, when A is a limit point for a set of planar points on l_A. Let the stationary tangent plane along l_A coincide with the plane xy and let the surface S in a sufficiently small neighbourhood of the point A be represented by the equation $z = z(x, y)$ over some circle $\Omega : x^2 + y^2 \leq R^2$, and let the point A have the coordinates $(0, 0, 0)$, and the generatrix l_A runs along the axis x. There are planar points on l_A arbitrarily close to A. We take one of them and designate it as $M(x_1, 0, 0)$, assuming $x_1 > 0$. As the point M is planar, and the tangent plane to S in M coincides with the plane xy, then on S through M we can draw the segment l_M, orthogonal to AM and totally lying in Ω. Let us take on l_A on the left from A some point $C \in \Omega$ and draw through the point C the vertical plane Π intersecting the segment l_M at some point N_1 inside the circle Ω. The segment CN_1 lies inside the circle Ω, and the cross section of the surface by the plane Π is some curve Γ projecting to the segment $\gamma = CN_1$. Let us show that part of the surface S for points (x, y) from the triangle $\Delta = CMN_1$ coincides with this triangle. Indeed, let us take an arbitrary point $Q \in S$ projecting to some internal point Q' of the triangle Δ. By the condition, some segment with the stationary tangent plane passes through the point Q. We designate as l_Q its maximum extension with preservation of the tangent plane. If this segment reaches one of the segments $[CM]$, $[MN_1] \subset S$, then the point Q will prove on the plane xy as on the common tangent plane, on which the segments $l_Q, [CM], [MN_1]$ are located. Let l_Q reach neither $[CM]$ nor $[MN_1]$ by any of its ends, then its projection $l_{Q'}$ to the plane xy intersects neither MN_1 nor CM. Let us draw the straight line CQ' on the plane xy. Let N' be a point of intersection of the straight line CQ' with the segment MN_1. The triangle CMN' has two sides lying on the surface S on its tangent plane xy. One of the ends of the segment $l_Q \subset S$ projects inside this triangle (or both its ends project to the side (CN'); herewith, all the projection – the segment $l_{Q'}$ – reaches neither C nor N'). Let Q_1 be such an end, Q_1' be its projection lying in the union of the open triangle CMN' and the interval (CN'). A segment l_{Q_1} passes through Q_1; along this segment the surface has the same tangent plane as at the point Q. Repeating the same reasoning with the segment l_{Q_1} as with the segment l_Q passing through the point Q, we will obtain that either the point Q_1 and, together with it, the point Q, lie on the plane xy (when the extension of the segment l_{Q_1} will meet the manifold $[CM] + [MN']$), or we should continue the constructions with respect to that of the ends Q_2 of the maximally extended segment l_{Q_1}, which is in the triangle CMN'', $N'' \in MN'$ constructed by analogy with the above construction. Continuing the similar constructions, we will either find after a finite number of steps that the point Q proves to be lying on the plane xy, or else we will obtain an infinite sequence of points $Q, Q_1, Q_2, \ldots$ on S, forming a connected chain of segments $[QQ_1], [Q_1Q_2], \ldots$, along which the surface has the same tangent plane. The projections of these segments form in the open triangle Ω a one-dimensional line, being simultaneously an open and a closed manifold with the monotone approximation to either the side CM (with the monotonicity of the ordinate) or to MN_1 (with the increase of the abscissa) or to both sides. Therefore, in the limit this chain of lines should come to the sides $[CM] + [MN']$; therewith, it will turn out that the point Q lies on the tangent plane coinciding with the plane xy, which shows the coincidence of the surface S with the triangle Δ. Let in the just performed constructions the ordinate y of the point N_1 be positive. Then we can perform similar constructions, taking inside the circle Ω on the segment l_M some point N_2 with the negative ordinate and obtain as the result that the entire triangle $\Delta = CN_1N_2$ lies on the surface S. Then the open triangle Δ will prove a plane neighbourhood of the point A, which is impossible. Thus, in this case we have a contradiction with the condition.

Consider now the case when the point A is not the limit point for the set of planar points on l_A. The intersection of the neighbourhood of the planar point $B \in l_A$ with l_A consists of planar points on l_A.

Among them, there is a non-planar point $D \in l_A, D \in (AB) \subset l_A$, which is the limit point for planar points from l_A. But above we have just proved that this is impossible.

Property 3.3. If a point A on a torse has no plane neighbourhood, then the segment passing through it, along which the tangent plane to the surface is constant, extends to both sides up to the boundary of the surface with the property of constancy of the tangent plane preserved.

For the proof, we need to assume the opposite: let the generatrix l_A during its extension come to a stop at some interior point $C \in S$. Then, through C, we can draw the generatrix l_C, along which the surface has the same tangent plane as along l_A. After that we can repeat verbatim the same reasoning as above (the role of the segment l_M will now be played by the segment l_C) and obtain that the point C is planar, which is impossible by the above proved property.

Thus, though in the definition of the torses we assumed that a rectilinear segment passes through each point of the surface along which the tangent plane to the surface is stationary, we said nothing of the expanse of that segment; now we know that for nonplanar points the segment extends to both sides, up to the edges of the surface.

Property 3.4. If some point A on S is not planar, then only *one* generatrix with the stationary tangent plane passes through it.

For the proof, let us assume the opposite and perform the previous reasoning for four angles between two generatrices.

Property 3.5. The proposition of Property 3.4 can be formulated thus: if more than one generatrix with the stationary tangent plane to the surface passes through some point, then this point has a plane neighbourhood on the surface.

In reality, for this property to be valid we should not request that the tangent plane to the torse be stationary along *both* generatrices, and it would be sufficient to assume its stationarity along only one generatrix (such a generatrix always exists by the definition of the

C^1-smooth torse). We shall not, however, prove this fact, as it will be obtained as a corollary of the below Theorem 3.2 and the following simple proposition.

Proposition 3.1. *If two generatrices pass through some point A on a C^1-smooth surface S with an l. E. metric, the point A is planar, i.e., has a plane neighbourhood on S.*

Indeed, if two generatrices pass through some point $A \in S$, both of them lie on the common tangent plane T_A. Then in a sufficiently small neighbourhood of the point A the arc of the shortest path connecting on S two arbitrary points on the two generatrices must lie on the same tangent plane T_A to S at point A. This is because, due to the isometricity of the neighbourhood of the point A on the surface to some circle on the plane T_A, the length of this shortest path is equal to the length of the rectilinear segment on T_A between the considered points, and thereby the shortest path on S coincides with this segment lying on the plane T_A. Note that for no generatrix did we assume the stationarity of the tangent plane along it.

Property 3.6. On a C^1-smooth torse, there can be no isolated generatrix with nonplanar points.

The proof is easily obtained by contradiction.

For some other properties of the torses, see Subsection 3.1.7.

3.1.5   The Burago class of C^1-smooth surfaces

Thus, C^1-smooth torses are, first, normal developable surfaces. But such surfaces have been proved ([124], pp. 695–696) to be locally isometric to the plane. Here we need to make some clarifications. The thing is that this book [124] has no concept of normal developable surfaces. But on pp. 694–695 it shows first that surfaces of zero extrinsic Pogorelov curvature have the structure exactly coinciding with that, which normal developable surfaces have (by definition!), and then it is proven that surfaces with *such a* structure are locally isometric to the plane. What is more, as noted by Yu.D. Burago, Pogorelov never makes use here of the fact that he considers surfaces of extrinsic curvature in his sense; he is based only on C^1 smoothness and on the structure of the surface as in the class of normal developable surfaces (note, by the way, that the properties of the generatrices used in the proof by Pogorelov – their uniqueness and extensibility up to the boundary – in our case are proved irrespective of Pogorelov's reasoning.) Second, torses are *saddle* surfaces, i.e., you cannot cut off the hump – a compact surface with the plane edge, which means that they are of *bounded extrinsic positive curvature in the Burago sense*[12] (BEPCB) [31, 32], but for such surfaces it has been proved that if the metric of zero curvature is immersed isometrically into E^3 as a C^1-smooth surface of class BEPCB, then such a surface is a normal developable surface [160]. Herewith, the following theorem is valid.

Theorem 3.2. *For a C^1-smooth surface to have an l. E. metric and simultaneously be a surface of bounded extrinsic positive curvature in the Burago sense, it is necessary and sufficient for it to be a C^1-smooth torse.*

Remark 3.4. The difference between Theorems 3.1 and 3.2 is not only that the class of BEPCB surfaces is broader, at least formally, than the class of BECP surfaces[13], but also that Theorem 3.1 does not assume *a priori* that torses are of BEPCB class. For greater clarity, we present a logical scheme of the propositions of the theorems.

Theorem 3.1   Theorem 3.2

(l.E.+BECP) $\Rightarrow$ torse   (l.E.+BEPCB) $\Rightarrow$ torse   •

(torse+BECP) $\Rightarrow$ l.E.   torse $\Rightarrow$ (l.E.+BEPCB)

3.1.6 C^1-smooth G-stable Shefel' surfaces

To the already known properties of torses, we add the following observation of fundamental importance: *the property of the metrics of these surfaces to be locally Euclidean does not change in affine transformations* (according to S.Z. Shefel' [158, 159], this property is

---

[12] This means that the function $m^1(\nu)$, assigning to each direction ν the number of points, at which it locally has a strictly supporting plane with the outward normal ν, is an integrable function on the sphere.

[13] Each BECP surface enters into the class of BEPCB surfaces, but the inverse is unknown, in particular, we do not know if all torses are BECP surfaces, but the fact that they are *a fortiori* in the BEPCB class was established above.

I.Kh. SABITOV

called *G stability* of the immersion of an l. E. metric in the form of normal developable surfaces or, otherwise, C^1-smooth torses, relative to group G of affine transformations of space). Indeed, in an affine transformation straight lines pass into straight lines, tangent planes pass into tangent planes, so a torse passes into a torse with local Euclidity of the metric respectively preserved.

What is more, it turns out that the property of G stability of the metrics of torses enables the assertion that *any* C^1-smooth immersion of an l. E. metric into E^3, G-stable relative to the group G of affine transformations, is a torse [32, 158]. As the result we have the *second characteristic* of C^1-smooth torses as C^1-smooth G-stable isometric immersions of l. E. metrics into E^3: *for an isometric immersion of an l. E. metric into E^3 in the form of a C^1-smooth surface to be G-stable relative to the group G of affine transformations of space, it is necessary and sufficient that this surface be a torse*[14].

3.1.7 *Some other properties of torses and open problems*

If a point M on some torse is planar, then, though an infinite set of rectilinear segments passes through it, *there can still be no* generatrix passing through it and coming by both endpoints to the boundary of the surface. An example (from [76]) is this: to each side of a flat triangle, let us attach cylindrical surfaces with the generatrices parallel to the respective side of the triangle (the smoothness of the glueing can, certainly, be of any regularity from $C^0 = C$ up to C^∞), and then a straight-line segment running on both sides up to the boundary of the surface could not be drawn through any point of the triangle. Still, if only two cylindrical domains are attached to the triangle, then exactly one generatrix running from edge to edge of the surface will be passing through each point of such a surface. But this property of the surface ceases to be true if it is reduced, i.e., if one considers separately any subdomain of the triangle as part of the initial surface.

This observation shows that generally speaking we can distinguish between the local and global definitions of the torse. The definition

---

[14] Exactly this characteristic property of torses was used in [160] mentioned above in connection with the proof of Theorem 3.2.

we operated with until now can be called *local* in the sense that it does not specify the length of the segment, which by condition passes through the considered point, and along which the surface has the same tangent plane. But in the *global* definition we request that *a straight-line segment, which extends to both sides up to the boundary of the surface with the stationary tangent plane to the surface along the entire extension, pass through each point of the surface.* Then in the above example the surface would not be a torse in the sense of its global definition. If we confine ourselves to the global definition of the torse, then the disposition and shape of plane domains on a non-analytical developable surface will not be so diverse as in the generally accepted definition, and this may lead to some new results pertaining to torses in the global sense, see, for example, further in Subsection 3.1.8. For this reason, it is natural to put the following question:

Open question 3.1. What can be said of the form of plane domains on a torse, if it is known that the unique generatrix, extended on both sides to the boundary of the surface, passes through each point of the surface?

With reference to the structure of torses, we can put yet other questions.

Open question 3.2. As it is known (see below in Subsection 3.2.5), in the class of smoothness C^2 torses without points of flattening can be surfaces of three types: 1) cylindrical; 2) conical; 3) made up from the tangents to some spatial curve. The question is whether *these three surfaces exhaust all kinds of torses in the class of smoothness* C^1? For the complete answer we would, probably, need to define an analogue of the concept of the point of flattening (e.g., as a point in which the tangent plane has a higher-than-first-order contact with the surface).●

The following question is a particular case of the previous question.

Open question 3.3. What should be the efficient definition of the line of striction for C^1-smooth torses?

Open question 3.4. Below we will see that the so-called *asymptotic parametrization* of a developable surface, i.e., the representation of the position vector $\mathbf{r}$ of a C^1-smooth torse S in the form

$$\mathbf{r}(u, v) = \mathbf{a}(u) + v\mathbf{b}(u), \tag{3.1}$$

where $\mathbf{a}(u)$ is the position vector of some directrix on the torse and $\mathbf{b}(u)$ is a continuous unit vector of the generatrices, yields no information on the real smoothness of a surface. What is more, if in the representation (3.1) we can assume *a priori* that the directing curve $\mathbf{r}(u)$ is C^1-smooth (for example, it can be considered to be the intersection $S \in C^1$ of the surface itself with some plane non-tangent to it), then the continuity of the vectors $\mathbf{b}(u)$ for the directions of the generatrices should, generally speaking, be either assumed or proved. For this reason, the **question** is posed as follows: what requirements to the vector functions $\mathbf{a}(u)$ and $\mathbf{b}(u)$ provide for the surface S with the representation (3.1) being in reality a C^1-smooth torse? Vice versa, let it be known that S is a C^1-smooth torse, then what necessary conditions are satisfied by the vector functions $\mathbf{a}(u)$ and $\mathbf{b}(u)$, which make it possible to check that a surface with the equation (3.1) is indeed in some coordinates (i.e., not necessarily in coordinates (u, v)) a C^1-smooth torse? In particular, how can we know when the tangents to a C^1-smooth spatial curve form a C^1-smooth developable surface? (For example, if the position vector of the generatrix has the second derivative, which at some point has a first-kind discontinuity, then the surface S formed by tangents to such a curve is obviously not C^1-smooth, because the normal to S has a discontinuity together with the second derivative of the position vector of the curve.) Apparently, these C^1-smooth curves, the tangents to which form a C^1-smooth surface, possess some geometrically interesting additional properties, e.g., the development of the tangents to them may prove to be a surface of class BEPCB not belonging to BECP surfaces.●

Open question 3.5. In Remark 3.3 we noted that for C^1-smooth torses of BECP class of surfaces the *a priori* weakening of conditions in the definition of a torse (substitution of "passage" through each point of a straight-line segment for "emanation from" each point of some segment also with the stationary tangent plane) in reality does not lead to broaden the set of torses. But if the assumption that the torses considered belong to the BECP class of surfaces is removed, then the author is unaware if the class of torses with "weakened" conditions (this class of surfaces can be called "semi-torses") remains coinciding with the class of torses with the original definition. Weakening of the conditions does not take "semi-torses" from the class of saddle surfaces, and, consequently, from the class of BEPCB surfaces, either; thus, as metric manifolds they remain in the class of MCB in the sense of Alexandrov (herewith, having a non-positive curvature) but it is not known if they would have an l. E. metric.●

If their metric will be locally Euclidean, then, on the strength of the results [160], "semi-torses" will also prove to be normal developable surfaces, i.e., nothing new will come out. If it appears that torses with "weakened" conditions do exist and do not coincide with the class of previous torses, then these surfaces will present an example showing that surfaces of class BEPCB are indeed broader than the BECP class of surfaces. Note that "semi-torses" also remain in their class under all affine transformations of space.

3.1.8 Structure of complete developable surfaces

We have already noted above the truth of the following:

Theorem 3.3. *A complete connected developable surface in E^3 is a cylinder.*

This theorem was proved by A.V. Pogorelov in 1956 and was included into his monograph [124][15]. The proof assumes that the surface has a smoothness C^1 and belongs to the class of BECP (bounded extrinsic curvature in the Pogorelov sense) surfaces. In [76], it was generalized for the case of hypersurfaces in any $E^n, n \geq 3$, with a complete l. E. metric but in the class of smoothness C^2. Later for surfaces in E^3 the works [108] and [172] proposed simpler proofs with reasoning only within the framework of elementary differential geometry. In

---

[15] The proof by Pogorelov, as by other authors, assumes the isometricity to the surface of the complete plane E^2 and thereby its orientability, which, certainly, does not affect the generality of the result, as it is always possible to pass to the universal covering surface.

[108], the surface was in fact required to be of smoothness no less than class C^4; and in [172], as is declared in the work itself, the smoothness was assumed to be of class C^2. Besides, instead of the unbounded extensibility of *any* geodesic, as is required in the definition of completeness by Hopf–Rinov, in [172] it is assumed that *no rectilinear generatrix on the surface contains any boundary points of the surface.* This means that the property of unbounded extensibility of geodesics is assumed to be satisfied only for geodesics in the form of straight lines, and in a weakened variant, too. It should only be satisfied for the generatrices passing through nonplanar points (from general theory, such generatrices are known to run from boundary to boundary of the surface or be extended unboundedly if there is no boundary); but for generatrices passing through a planar point and not coinciding with a complete straight line, it is only assumed that if they do not go as far as the edge of the surface but are ended at some interior point of it, then they are further extended already as some "curvilinear" geodesics without specifying the limits of their extension. Surfaces belonging to this class of surfaces are, for example, such incomplete surfaces in the classical sense as an infinite circular cylinder, in which one generatrix or even the whole band between two generatrices is removed. But the condition of C^2 smoothness of the surface declared by the author of [172] is in fact insufficient for his reasoning, because, for example, on p. 629 of his paper he makes use of the derivative $(k_1)_v$, where k_1 is the principal curvature not equal to zero; besides, it is not clear why under conditions of C^2 smoothness of the surface it can be guaranteed that only one line of the curvature can pass through one point in a given principal direction, see p. 166 in [80][16]. The proof of the theorem itself does not seem to be impeccable, either, as on p. 634 in [172] it is asserted without sufficient grounds that the surface band (projected onto the plane (x, y)) along the generatrix receding into infinity does not shrink to the zero width. This proposition would have been justified if we were talking about the *internal* width of the band, but for the assessment of the *external* width of the band we need to know the uniform estimates

---

[16] These lapses, however, do not affect the validity of the auxiliary propositions on the properties of the generatrices proved [172], as they are really true in the class C^2 and were proved above in Subsection 3.1.4 even in the class C^1.

of variation of the normal to the surface in the direction orthogonal to the generatrix and the uniform estimates of the external width of the investigated band depending on the variation of the normal. Nevertheless, we will prove the theorem *as proved by Stoker, and, besides, assuming only that C^1-smooth torses are considered,* on which the generatrices are known to possess Properties 3.1–3.6, which were proved earlier.

Thus, *let us be given a C^1-smooth developable surface S in the form of a torse, complete in the sense of [172] and not coinciding with the plane.* Such a surface will definitely have a nonplanar point p_0, through which the (even locally) unique generatrix l_0, also consisting of nonplanar points, passes. By Property 3.3, this generatrix extends to the boundary of the surface, and as it is impossible by the condition (or because there is no boundary), it is situated on the surface as a complete straight line. Assume first that the straight line l_0 is the boundary of some connected domain D of planar points, which domain should be situated on the plane T_0 tangent to the surface. Let a half-plane containing D be designated as T^+. Let D be not coincident with T^+ (if D coincides with T^+, we will perform the below reasoning for another half-plane T^-). The relative boundary D has points different from points on l_0 (otherwise the straight line l_0 cuts off from S the part coinciding with the plane domain D). Let M be such a point. By Property 3.1, some unique generatrix l_1 passes through M; the generatrix belongs to the relative boundary D and is, by the condition of the theorem, a complete straight line. This straight line is on the same plane as l_0 and cannot intersect with l_0, consequently, it is parallel to it. In the band between l_0 and l_1 the domain D cannot have boundary points. Indeed, let us have a boundary point N not belonging to l_0. Then the unique generatrix l also passes through that boundary point; it is also parallel to l_0 and consists of nonplanar points. But this implies that the straight line l_0 is separated from the straight line l_1 by the generatrix l, which contradicts the connection of the domain D with respect to the points in the neighbourhoods of the straight lines l_0 and l_1. Thereby, for the moment we have that *if there is a connected plane domain on the surface S, it represents a band between two parallel lines and the unique unbounded generatrix passes through each point of this plane domain, namely the one, which is parallel to the edges of the band.*

Now assume that any neighbourhood of the straight line l_0 has nonplanar points. Consider the image of this neighbourhood on the development S to the plane. Let the axis y correspond to the generatrix l_0. From each point of the axis y, orthogonally to the axis towards $x > 0$, draw a geodesic, i.e., a segment of the straight line, extending it as far as it is possible. Let the lengths of the maximally extended geodesics be equal to $L(y) > 0$. If $L(y) \geq \varepsilon > 0$ for all y, then the generatrix going out of any nonplanar point $(x_0, 0), 0 < x_0 < \varepsilon$ (certainly, on the plane (x, y) we are talking of the pre-images of the generatrices for the development of the surface onto the plane) will be parallel to l_0, otherwise it will have to intersect l_0, which is impossible. As we established above, in the band corresponding to the connected domain of planar points, in each point there is also a unique generatrix extending as a complete straight line and parallel to the sides of the band. This straight line cannot intersect l_0, either, so the band will be stratified into generatrices parallel to the generatrix l_0.

But if $\inf L(y) = 0$, our reasoning is as follows. We introduce the function

$$l(y) = \begin{cases} \inf_{0 \leq \eta < y} L(\eta), & y \geq 0, \\ \inf_{y < \eta \leq 0} L(\eta), & y \leq 0. \end{cases}$$

This function is non-ascending during the increase of $|y|$ and piecewise continuous. Its graph γ with added "steps" (the "steps" will be at the discontinuity points of the function $l(y)$) and the axis y bound a simply connected domain $\Omega : 0 < x < l(y)$. The curve γ consists of three types of arcs: 1) horizontal steps – segments of the form $a \leq x \leq b, y = c$, where c is the discontinuity point of the function $l(y)$; what is more, the point (a, c) corresponds to the boundary point of the surface S; 2) the vertical open segments of the form $x = a, c_1 < |y| < c_2$, whose points correspond to interior points of the surface; 3) continuous arcs of the monotonously decreasing curves $x = L(y)$ whose points correspond to the boundary points of the surface S (Figure 1).

We will now leave out the generatrices[17] from points of the interval $(0 < x < x_0, y = 0)$, but first we show that their

---

[17] In reality, in this and the next paragraph we talk about the pre-images of the generatrices on the plane (x, y).

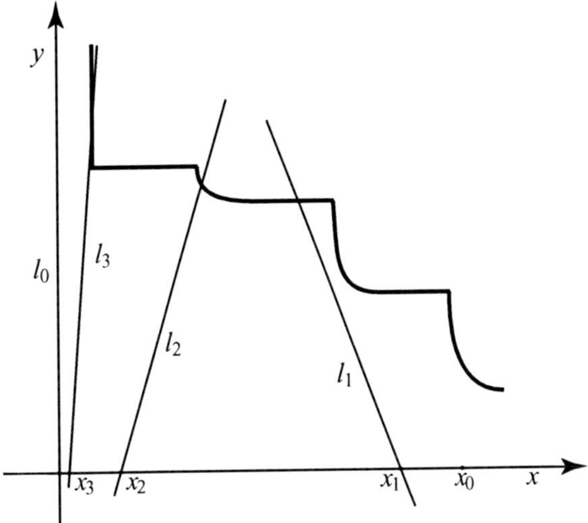

Figure 1

directions at sufficiently small x_0 form a continuous family. Let the point $(0,0)$ correspond to some point $M_0 \in l_0 \subset S$ and let some neighbourhood of the point M_0 develop isometrically onto the circle $\omega_0 : x^2 + y^2 < r_0^2$ with the condition $\omega_0 \subset \Omega$. The generatrices l going out of the points of the interval $(0 < x < x_0 = r_0)$ cannot intersect the vertical diameter $x = 0, -r_0 < y < r_0$, of the circle, corresponding to the gencratrix l_0. Hence, it follows that the generatrix $l(\xi)$ going out of the point $(\xi, 0), 0 < \xi < x_0$, intersects the circumference $x^2 + y^2 = r_0^2$ at a point with polar coordinates $r_0, \cos\theta = 2\xi r_0/(\xi^2 + r_0^2), \sin\theta = (r_0^2 - \xi^2)/(\xi^2 + r_0^2)$. Therefore, the directions of these generatrices at $\xi \to 0+$ tend to the axis y, i.e., the directions of the generatrices are continuous on the right-hand side in the point $\xi = 0$. Their continuity on the left-hand side in the point $\xi = 0$, as well as the continuity in the other points ξ are proved similarly.

Let us follow the generatrix $l(\xi)$ going out of some point $(\xi, 0)$. If in its unbounded extension it would not intersect with the generatrix l_0, then it would be parallel to it. But if there is an intersection, then before intersecting with l_0 the generatrix $l(\xi)$ should go outside the bounds of the domain Ω, because generatrices cannot intersect within a simply connected domain. Consider the first intersection of $l(\xi)$ with γ – the boundary of the domain Ω. It cannot occur at points of third-type arcs, as they are boundary points of the surface S. If it did occur at some point (a, c) in a first-type arc, then we will follow new generatrices by continuously shifting to the left along the segment $(\xi, 0)$. The rectilinear segments $(0 \leq x \leq \xi, y = 0), (x = 0, 0 \leq y \leq c), (0 \leq x \leq a, y = c)$ and the segment of the generatrix $l(\xi)$ from the point $(\xi, 0)$ up to the point (a, c) bound the trapezium T totally lying inside Ω. Therefore, the generatrices $l(x), x < \xi$, cannot intersect with either $l(\xi)$ or l_0 within this trapezium. Assume that the intersection point (a, c) of the generatrix $l(\xi)$ with γ is assigned to a first-type arc. Then during the shift to the left along the interval $(\xi, 0,)$ the generatrices will also continuously be shifting to the left, intersecting γ at points of the same first-type arc, and finally these intersection points will come to the left end of the horizontal step $y = c$, which is a boundary point of the surface S, which is impossible by the assumed structure of the surface. Now if the intersection point is a point of a second-type arc, then the points of intersection of the new generatrices with γ will shift upwards along this region and will arrive at the vertex of the vertical segment; the vertex is also a boundary point of the surface S. Again we have a contradiction, and thereby it is proved that all generatrices emanating from points of the interval $(x_0, 0)$ are parallel to the generatrix l_0 and fill the band $0 \leq x \leq x_0$.

Let us now consider the respective generatrices on the surface itself. Let us show, following [124], that they are also parallel between themselves. For this, it is sufficient to show that they cannot be divergent lines. Indeed, the pre-images of any pair of generatrices on the plane (x, y) are parallel and therefore they have pairs of points receding into infinity and preserving the distance d between them. The images of these points on the surface also have an intrinsic distance not exceeding d. However, if the lines are divergent, as they recede into infinity, the spatial distance between them increases infinitely, which is impossible, because it is not greater than the intrinsic distance. This contradiction shows that the generatrices are parallel between themselves, which is what was required to be proved.

3.1.9 One more theorem on complete developable surfaces

The Pogorelov theorem on the structure of a complete surface with an l. E. metric admits the following generalization obtained by A.A. Borisenko in [23] (see also [26]) and representing the "exterior" analogue of the "intrinsic" theorem by Toponogov from [178].

Theorem 3.4. *Let F be a two-dimensional surface of class C^0 in $E^n, n \geq 3$, isometric to the plane and containing the straight line p of the ambient space. Then F will be a cylindrical surface with the generatrices parallel to the straight line p.*

In Section 3.1 in Remark 3.1 we noted that without assuming the smoothness of the surface the local Euclidity of its metric is understood as the coincidence of the lengths of corresponding curves on the pre-image (in this case, on the plane) and on the surface itself. In this theorem, exactly this understanding of the isometricity of the plane of the surface F, given as a continuous image of the plane, is borne in mind.

Let us adduce the proof of the theorem from [23], as it is not very long and rather elegant, though it requires some clarification. Denote as $\bar{F}$ the plane isometric to the surface F. All objects on the plane $\bar{F}$ and on the surface F, corresponding one to another by isometry, will be denoted by the same symbols with the bar added, when they pertain to the plane.

Let $\bar{p}$ be a straight line on the plane $\bar{F}$, corresponding to the straight line p on the surface F. On the straight lines $\bar{p}$ and p let us introduce a general natural parameter s, counted off from some points $\bar{P}(0) \in \bar{p}$ and $P(0) \in p$, respectively. Denote as $\bar{M}(s)$ the shortest arc (i.e., the straight line) on $\bar{F}$, passing perpendicular to $\bar{p}$ at the point $\bar{P}(s) \in \bar{p}$. Show that for each s the intersection $F(s)$ of the surface F with the hyperplane $E^{n-1}(s)$, drawn through the point $P(s) \in p$ orthogonally to p, coincides with $M(s)$ – the image of $\bar{M}(s)$ on F.

Indeed, first note that the straight lines $\bar{M}(s)$ on the plane $\bar{F}$ are parallel and therefore do not intersect. Each point P on the surface F is assigned to some line $M(s)$. Assume that for some value of $s = s_0$ a point $Q \in M(s_1), s_1 \neq s_0$ is contained in $F(s_0) = E^{n-1}(s_0) \cap F$. Let $\bar{Q}$ be a point on the plane $\bar{F}$, corresponding to Q and assigned to $\bar{M}(s_1)$ (we refine the choice of the point $\bar{Q}$, because the surface F may have self-intersections, and one of the pre-images may lie on $\bar{M}(s_0)$). Consider on $\bar{F}$ a triangle $\bar{P}(s_0)\bar{Q}\bar{P}(s)$ with as yet arbitrary $s \neq s_0$. As $\bar{Q} \in \bar{M}(s_1)$ and $\bar{M}(s_1) \cap \bar{M}(s_0) = \emptyset$, then the shortest arc $\bar{P}(s_0)\bar{Q}$ is not orthogonal to $\bar{p}$. We choose a point $\bar{P}(s) \in \bar{p}$ on the ray, which forms an acute angle with the shortest arc $\bar{P}(s_0)\bar{Q}$. At a sufficiently large $|s|$ the angle $\bar{p}(s_0)\bar{Q}\bar{P}(s)$ will be obtuse, so for the lengths of the sides of the considered triangle we will have an inequality

$$|\bar{P}(s_0)\bar{P}(s)| > |\bar{Q}\bar{P}(s)|. \tag{3.2}$$

In the spatial triangle $P(s_0)QP(s)$ the angle at the vertex $P(s_0)$ is right, hence $|P(s)Q| > |P(s_0)P(s)| = |\bar{P}(s_0)\bar{P}(s)|$, and $|QP(s)| \leq |\bar{Q}\bar{P}(s)|$. We have that $|\bar{P}(s)\bar{Q}| > \bar{P}(s_0)\bar{P}(s)|$, which contradicts the inequality (3.2). Therefore, $E^{n-1}(s_0)$ can have no points from $M(s_1)$ at any $s_1 \neq s_0$. Similarly, $M(s_0)$ can have no common points with any $E^{n-1}(s_1)$, $s_1 \neq s_0$, therefore,

$$F(s) = M(s). \tag{3.3}$$

Further, let $\bar{Q}$ be an arbitrary point at $\bar{F}$, not lying on the straight line $\bar{p}$. We draw through $\bar{Q}$ a straight line $\bar{q}$ parallel to $\bar{p}$ and show that its image on F is a straight line in E^n, parallel to p. Take on $\bar{q}$ two arbitrary points $\bar{Q}_1$ and $\bar{Q}_2$ and drop perpendiculars from them to $\bar{p}$ with the bases in $\bar{P}_1, \bar{P}_2 \in \bar{p}$. Consider the spatial quadrangle $P_1P_2Q_1Q_2$. Span some Euclidean space E^3 (unique if these four points are in general position) over this quadrangle. Introduce in this E^3 a rectangular system of coordinates with the origin at P_1 and with the axis z along the straight line p. Then, with account for the equality (3.3), the points P_1, P_2, Q_1, Q_2 will have the coordinates $(0,0,0)$, $(0,0,z_2)$, $(x_1,y_1,0)$, (x_2,y_2,z_2) respectively with distances

$$|P_1P_2| = |z_2|, \quad |Q_1Q_2| = \sqrt{(x_2 - x_1)^2 + (y_2 - y_1)^2 + z_2^2} \geq |P_1P_2|.$$

But $|P_1P_2| = |\bar{P}_1\bar{P}_2| = |\bar{Q}_1\bar{Q}_2| \geq |Q_1Q_2|$, hence, $|P_1P_2| = |Q_1Q_2|$. But this is possible only at $x_1 = x_2, y_1 = y_2$, i.e., the segment Q_1Q_2 is parallel to the straight line p and, by the arbitrariness of choice of the points $\bar{Q}_1$ and Q_2 it lies on the surface F. The theorem is proved.

3.1.10 Surfaces of class $C^{1,\alpha}$

As we have already noted above, in the classes of regularity $C^{1,\alpha}$, $\alpha <$ 1/7, there are surfaces, which are obtained by bending the plane and contain no rectilinear generatrices. But if we are to search for isometric immersions of l. E. metrics in stronger classes of regularity, then the structure of these surfaces in the form of a torse is obtained without any additional requirements of their assignment to BECP or BEPCB classes. In this respect, there is a result [159], which refines the corollary of Theorem 3.1.

Theorem 3.5. *For the surface of class $S \in C^{1,\alpha}, \alpha > 2/3$, in E^3 to have a locally Euclidean metric, it is necessary and sufficient that it is a torse. Herewith, the torse will of necessity be a BECP surface.*

The proof of this theorem is based on very difficult results by Yu.F. Borisov on the coincidence of extrinsic and intrinsic curvatures of the surface of class $C^{1,\alpha}$, $\alpha > 2/3$, see [27]. A recently published work [29] presents an unpublished result by S.Z. Shefel' affirming that $C^{1,2/3}$-smooth surfaces with a regular metric of fixed-sign curvature are also surfaces of bounded extrinsic curvature in the Pogorelov sense, so that the theorem also extends to surfaces of this class of smoothness.

3.1.11 On the "development" of surfaces

The widely used term "developable surfaces" in its day-to-day meaning implies that a surface can be unfolded or put onto a plane without folds and bends, as a carpet is put onto the floor. This is a visual illustration of the isometricity of a developable surface S to some domain D corresponding to S on the plane by isometricity (which then, by the way, is called a *development* of the surface S). But the theory of isometric transformations distinguishes between the concepts of "local" and "global" isometricity (or isometricity "in the small" and isometricity "in the large"), the same way as the concepts of "discrete" and "continuous" isometric transformation. The latter type of isometric transformations is called *bending*, and two surfaces joined by bending are said to be *superposable* onto each other (by bending). Besides, we can speak of developing onto a plane so that there is a bijective correspondence between points of the surface and its development onto the plane, and we can admit the "overlaps" of the development. For this reason, when talking of "developing" a developable surface onto a plane, it is necessary in general to specify which form of "development" is meant. For instance, a truncated cone or a part of a circular cylinder homeomorphic to a ring or a flat Möbius band do not develop onto a plane "in the large", though locally it can be done. In Subsection 4.2.3, we will construct an embedding of a doubly covered circular ring into space in the form of a cylindrical surface, for which an isometry for this ring can be said to exist, but it is as yet unknown if it can be embedded onto the ring by bending. Thus, the issue of *how* isometry of a torse onto a surface occurs always requires separate consideration.

3.2　*External structure in smoothness classes* C^2 *and higher*

A general description of the geometry of C^n-smooth, $n \geq 2$, developable surfaces is given by Theorem 3.6, from which it follows that in the analytical class the entire surface is either a domain on the plane or a cylinder or a cone or consists of tangents to some spatial curve. In Russian-speaking literature just these latter surfaces are usually called "torses". In the well-known educational monograph by V.F. Kagan ([94], §30), though, the term "torse" is used as a general name for all developable surfaces, and for a surface from tangents to a spatial curve the term 'tangential torse" is used, so we will also use this term and sometimes, by analogy with the terminology of English-speaking literature, will use the term "the development of tangents" for such surfaces.

In a non-analytical case, a general developable surface can consist of the union of all given kinds of surfaces.

3.2.1　*Local structure of developable surfaces*

In class C^2 and higher, developable surfaces admit, at least locally, the representation as a solution $z = z(x, y) \in C^2$ of the so-called *trivial* Monge–Ampère equation

$$z_{xx} z_{yy} - z_{xy}^2 = 0, \tag{3.4}$$

from which many new properties of these surfaces can be derived. These solutions have been described most fully in [181, 183] (see also pp. 769–770 in [77]). One of the results obtained in [181] and [183] will be presented here in the following formulation slightly more general than in those works:

Theorem 3.6. *Let $z(x, y) \in C^n$, $n \geq 2$, be a solution of the equation (3.4) in some domain D. Then, in the assumption that $z_{xx} \neq 0$, the relation $z_{xy} : z_{xx}$ is C^{n-1}-smooth, without the obligatory assignment of $z(x, y)$ to class C^{n+1}; and of the relation $z_{xy} : z_{xx}$, to C^n. Besides, the surface $z = z(x, y)$ will be either a cylinder or a cone or a torse, or else will consist of a union of torses with cones and/or cylinders[18].*

With respect to smoothness, the theorem can be commented thus: unlike elliptic equations for solutions of the parabolic Monge–Ampère equation there is no automatic increase of the initial smoothness, but there is only an improvement of smoothness by one order for the relation $z_{xy} : z_{xx}$.

Proof. (1) On some segment $[u_1, u_2]$, $u_1 < u_2$, let a function $g(u) \in C^{n-1}$ with the condition $g'(u) \neq 0$ and a function $f(u) \in C^n$, $n \geq 2$, be given. Following [183], we introduce three functions

$$x(u, v) = g(u) - vf'(u)$$

$$y(u, v) = v \qquad (3.5)$$

$$z(u, v) = ux(u, v) + vf(u) - G(u), \ G(u) = \int_{u_0}^{u} g(t)dt,$$

given in the band $\Pi : \ u_1 \leq u \leq u_2, -\infty < v < +\infty$. These three functions determine some parametrically given surface S with the position vector $\mathbf{r} = \mathbf{r}(u, v) = \{x(u, v), y(u, v), z(u, v)\}$, on which the vector product is:

$$[\mathbf{r}_u, \mathbf{r}_v] = \{u, -f, 1\}x_u. \qquad (3.6)$$

Therefore, the surface S is regular everywhere besides the points of the curve $L : x_u = g'(u) - vf''(u) = 0$, which (if it exists) splits

---

[18] In courses of differential geometry, the structure of developable surfaces is usually expounded in the assumption of at least the C^3 smoothness of the surface or without specifying the requirements for the smoothness of the surface at all, so the main aim of this point is to establish the known facts of the local geometry of developable surfaces at the minimal classical condition of C^2 smoothness.

the band Π into two domains $D^+ : g'(u) - vf''(u) > 0$ and $D^- :$ $g'(u) - vf''(u) < 0$. Let us study the problem of the representability of the surface in the form $z = z(x, y)$. The mapping

$$x = x(u, v) = g(u) - vf'(u), \quad y = y(u, v) = v \qquad (3.7)$$

has in the band Π the Jacobian

$$J(u, v) = x'_u = g'(u) - vf''(u) \neq 0. \qquad (3.8)$$

This means that in each of the above introduced domains D^+ and D^- (or in the entire band, if there is no curve L) the Jacobian is of constant sign, so the corresponding surface $S^\pm$ for each of the domains $D^\pm$ is a regular surface with locally single-valued projectability to the plane xy, but at points of the curve L the surface S even locally may have no one-to-one projection in the sense that in Π there is no neighbourhood U of the point $(u_0, v_0) \in L$, for which the surface S^+ is represented as $z = z(x, y)$ with the single-valued projection to the image of the domain $U \cup D^+$.

Let for definiteness $g'(u) > 0$. Consider the domain D^+ containing a segment of the straight line $v = 0$, the image of which in the mapping (3.7) is the segment $y = 0$, $g(u_1) \leq x \leq g(u_2)$, of the axis x. In the domain D^+ the function $y(u, v)$ is monotonous along the straight lines $u = const$, and $x(u, v)$, by (3.8), is monotonous along the intervals of the straight lines $v = const$ lying in D^+. The entire domain D^+ is connected, but the curve L in the general case may consist of several or even a countable set of connected components projected one-to-one to the axis u. In particular, the curve L contains no points altogether with coordinates (u, v), for which $f''(u) = 0$. If the curve L defined on the whole over the segment $[u_1, u_2]$ is continuous and monotonous, then the mapping (3.7) diffeomorphically translates each of the domains $D^\pm$ to some domain $D^\pm_*$ on the plane xy, and each of the surfaces $S^\pm$ will have a one-to-one representation $z = z^\pm(x, y)$. If L is convex upwards (downwards) and is situated in the domain $v > 0$ ($v < 0$), then the diffeomorphism is guaranteed only for the domain D^+. In the general case, in the absence of diffeomorphism on D^+ in the large, we are talking of the local representability of the surface in the form $z = z(x, y)$ over some domain D^+_*. For instance, if everywhere $f''(u) < 0$, then the curve L will be located in the domain $v < 0$ and the entire domain $v \geq 0$ belongs to D^+, and the mapping (3.7) is diffeomorphic in it. And at $f''(u) > 0$ the diffeomorphism (3.7) is guaranteed in the domain $v \leq min \dfrac{g'(u)}{f''(u)}$, $u_1 \leq u \leq u_2$. But in general the problem of the maximal domain of diffeomorphism of the mapping (3.7) appears to be not very simple. A simple example with the functions $g(u) = u + u^3$, $f'(u) = u$, $-1 \leq u \leq 1$, shows that the apparently expected result – the entire domain D^+ – proves wrong: in it, on any segment $v = C, 1 < C < 2$, there are pairs of points belonging to D^+ and disturbing the injectivity of the mapping. The most "nice" symmetrical domain of diffeomorphism is the infinite half-band $u_1 \leq u \leq u_2, -\infty < v < 1 + u^2$. The answer of the same kind also remains valid for the case of the infinite domain $-\infty < u < +\infty$, for which the corresponding surface S of the form $z = z(x, y)$ proves to be defined on the entire plane xy with the removed ray $x = 0, 1 \leq y < +\infty$. There are also other variants of constructing such solutions of the equation (3.4). In this connection, we can put a more general question:

Open question 3.6. How can we describe the maximal domains of existence for distinct-from-cylinders surfaces $z = z(x, y) \in C^n, n \geq 2$, considered as the solutions of the equation (3.4), in particular, represented in the form (3.5)? For instance, for surfaces of the type of a cone there is a surface $z = \sqrt{x^2 + y^2}$, defined over the entire plane with one punctured point. Now we saw that for torse-type surfaces there is a surface $z = z(x, y)$ defined over the entire plane with a removed rectilinear or curvilinear ray. Arc there solutions of the equation (3.4), defined over the entire plane with one removed segment, with some removed points, segments, rays etc.? How, using the solution known in the neighbourhood of some point, can we determine its maximal domain of existence, determine whether there are bounded domains among them, whether there are conditions of existence/non-existence of the solution of the Dirichlet problem or Cauchy problem for this equation, etc.? Certainly, when doing this, we can impose various additional conditions on the solutions, e.g., assume a certain smoothness, the execution of inequalities $z_{xx} \neq 0$ or $z_{xx}^2 + z_{yy}^2 \neq 0$ or admit the existence of different-structure parts of the surface etc. In this connection, recall the works [54, 82, 185] and [111], which have similar statements of the problems for surfaces of negative or non-positive curvature. The reader can also see a new work [5] with the statement of the Cauchy problem for surfaces with flat metrics in the sphere S^3.$\bullet$

We continue to study the considered surfaces. By straightforward calculations, we can see for ourselves that

$$u_x(x,y) = \frac{1}{g'(u) - vf''(u)}, u_y(x,y) = \frac{f'(u)}{g'(u) - vf''(u)}, v_x = 0, v_y = 1,$$

i.e., $u(x,y) \in C^{n-1}(D_*^+), v(x,y)$ is analytical. Besides,

$$z_u = ux_u(u,v), \; z_v = f(u) - uf'(u),$$

$$z_x(x,y) = u(x,y), \; z_y(x,y) = f(u(x,y)),$$

therefore, $z(x,y) \in C^n(D_*^+)$. From the same formulae, we have the following important geometrical conclusion: *straight lines (correspondingly, rays or segments)* $u = const$ *located in the domain* $D^+ \subset \Pi$, *and only they, pass into generatrices of the surface S, whose projections to the plane xy have the equations $x = g(u) - yf'(u)$ with respective variation limits of x and y.* The directions of the generatrices are given by the vectors

$$\mathbf{l} = \{-f'(u), 1, f(u) - uf'(u)\}. \tag{3.9}$$

Further, we have

$$z_{xx} = \frac{1}{A} \neq 0, \; z_{xy} = \frac{f'(u)}{A}, \; z_{yy} = \frac{f'^2(u)}{A}, \; A = g'(u) - vf''(u),$$
$$\tag{3.10}$$

whereby the composite function $z(x,y)$ determined by (3.5) satisfies the equation (3.4). The C^n-smoothness of this function is exact in the sense that if the functions $g(u)$ and $f(u)$ have no greater smoothness than that specified in their choice, then the solution $z(x,y)$ of the equation (3.4) constructed from them is not in the general case of class C^{n+1}. On the other hand, the relation $z_{xy}(x,y) : z_{xx}(x,y) = f'(u(x,y)) \in C^{n-1}(D_*^+)$ and again in the general case it can not be asserted to have the smoothness greater than C^{n-1}.

(2) With respect to smoothness, we have proved the theorem yet in the assumption that the solution of the equation is presented in a parametric form (3.5). Now let $z(x,y) \in C^n, n \geq 2$, be an arbitrary solution of the equation (3.4) in some domain of the plane xy, satisfying in this domain the condition $z_{xx} \neq 0$. Let us show that in this case the relation $z_{xy}(x,y) : z_{xx}(x,y)$ is of class C^{n-1}, too. Herewith, we will use a somewhat different course of reasoning than in [181], where this fact was proved for $n = 2$, because its proof in [181] has one unjustified proposition: namely, the establishment of smoothness of the relation $z_{xy}(x,y) : z_{xx}(x,y)$ is reduced to the verification of smoothness of the relation $z_{xy}(x,0) : z_{xx}(x,0)$. The matter is that the projection of the generatrix of the surface S : $z = z(x,y)$ onto the plane xy, passing through the point (x,y), and along which the relation $z_{xy} : z_{xx}$ is constant, intersects the axis x at some point $(\xi(x,y),0)$, so $z_{xy}(x,y) : z_{xx}(x,y) = z_{xy}(\xi(x,y),0) :$ $z_{xx}(\xi(x,y),0)$, but the C^1-smoothness of the function $\xi(x,y)$ in the work [181] in this place was not proved. Therefore, we will first show that the result on the existence of asymptotic parametrization

$$\mathbf{r} = \mathbf{a}(u) + v\mathbf{b}(u), \tag{3.11}$$

established in [80] for the position vector of a C^2-smooth surface of zero curvature without points of flattening, in which the direction vectors $\mathbf{b}(u)$ of the generatrices and the position vector $\mathbf{a}(u)$ of the directrix γ are of smoothness class C^1, is also valid in the general case in the sense that *if a surface without points of flattening is of smoothness $C^n, n \geq 2$, then in asymptotic parametrization (3.11) the vectors $\mathbf{b}(u)$ and $\mathbf{a}(u)$ have the smoothness of class C^{n-1}.* Indeed, on the surface S : $z = z(x,y)$ along the generatrices the tangent plane is constant, therefore on both of them the functions $p = z_x(x,y)$ and $q = z_y(x,y)$ are constant. As by the condition $z_{xx} \neq 0$, then the generatrices are not parallel to the plane xz, i.e., their projection to the axis y is non-zero. We make the substitution of the variables $u = p(x,y), v = y$, which is locally invertible with $y = v, x = x(u,v) \in C^{n-1}$. In the new variables, the position vector $\mathbf{r}$ has the form $\mathbf{r} = \{x(u,v), v, z(u,v)\} \in C^{n-1}$. Each coordinate line $u = const$ on the surface is assigned some generatrix, so along it the tangent vector $\mathbf{r}_v = \{x_v, 1, z_v\}$ is constant. This means that the functions x_v and z_v depend only on u. Putting $x_v = h(u), z_v =$

$H(u)$, we obtain the representations $x = x(u, v) = vh(u) + h_1(u) \in C^{n-1}$, $z = z(u, v) = vH(u) + H_1(u) \in C^{n-1}$. Therefore, the functions $h(u), h_1(u), H(u), H_1(u)$ are of class C^{n-1}, and we obtain asymptotic parametrization $\mathbf{r} = \{h_1(u), 0, H_1(u)\} + v\{h(u), 1, H(u)\}$ of required smoothness.

Let us turn again to the substitution of the variables $u = z_x(x, y)$, $v = y$, $x = x(u, v)$, $y = v$. Now we know that $x_v(u, v) = h(u) \in C^{n-1}$. Therefore, differentiating the equality $u = z_x(x, y)$ with respect to v, we obtain $0 = z_{xx}x_v + z_{xy}$, from which we have

$$z_{xy}(x, y) : z_{xx}(x, y) = -x_v(u(x, y), y) \in C^{n-1},$$

which is what was required to be proved.

(3) In addition to the propositions of the theorem, we will show that the representation of the solution of the equation (3.4) in the form of (3.5) is general in the sense that *any* solution of (3.4) is presentable in the form of (3.5) at least locally. For this we will again use one improved argument from [183]. Let $z(x, y) \in C^n, n \geq 2$, be some solution of the equation (3.4) with the condition $z_{xx} \neq 0$. Using this condition, we rewrite the equation (3.4) as $(z_y)_y = \dfrac{z_{xy}}{z_{xx}}(z_y)_x$. From the previous part **(2)** of the proof we know that the relation $z_{xy} : z_{xx}$ is of C^{n-1}. Consider the linear equation

$$q_y = \frac{z_{xy}}{z_{xx}} q_x \tag{3.12}$$

with the sought-for function $q(x, y)$ and the C^{n-1}-smooth coefficient. Assume $u = z_x(x, 0) = g^{-1}(x) \in C^{n-1}$, $f(u) = z_y(x, 0) = f(g^{-1}(x))$. As $z_{xx} \neq 0$, then there is the inverse function $x = g(u) \in C^{n-1}$; herewith, $g'(u) = z_{xx}(x, 0) \neq 0$; besides, $f'(u) = z_{xy}x_u = z_{xy} : z_{xx} \in C^{n-1}$, therefore, $f(u) \in C^n$. Using the formula (3.5), let us construct by the functions $g(u)$ and $f(u)$ the solution $\tilde{z}(x, y)$ of the equation (3.4). On the strength of the formulae (3.10), for this solution the function $\tilde{q}(x, y) = \tilde{z}_y$ satisfies the equation (3.12) with the Cauchy condition $\tilde{q}(x, 0) = f(g^{-1}(x))$. But the initial considered solution $z(x, y)$ of the equation (3.4) also yields the function $q = z_x(x, y)$ satisfying the same equation with the same Cauchy condition. Due to the uniqueness of the solution of the Cauchy problem at the C^1-smooth coefficient of the equation (3.4), we obtain that $z_y = q =$

$\tilde{q} = \tilde{z}_y = z_y$. In turn, the solution of the equation $w_y = q(x, y)$ with the Cauchy condition $w(x, 0) = \int_0^x g^{-1}(t)dt$ is also unique; because of this, its two solutions $w = z$ $w = \tilde{z}$ coincide, i.e., we have obtained that the arbitrarily taken solution $z(x, y)$ of the equation can be represented in the parametric form (3.5).

(4) Now we will show that in the class of smoothness C^2 the surface without points of flattening can have only three types of structure – a cylinder, cone or development of tangents to some spatial curve. Here our exposition will also be done along the lines of [183] and also with some required clarifications.

As we established in part (1) of the proof, the solution of the equation (3.4) exists in the domain D_*^+ of the plane xy, which is an image of the domain $D^+ : u_1 < u < u_2, g'(u) - vf''(u) > 0$ (in the assumption that $g'(u) > 0$) in the mapping of (3.7). In the domain $D^- *$ the surface $S^- : z = z(x, y)$ can be also presented as a ruled surface

$$\mathbf{r} = \{x, y, z\} = \mathbf{a}(u) + v\mathbf{b}(u) = \{g(u), 0, ug - \int g\} + v\{-f', 1, f - uf'\},$$
$$(3.13)$$

on which points of constant value u set rectilinear generatrices and which can be not connected because of the non-connectivity of the curve $L : g'(u) - vf''(u) = 0$. The parametric formulae (3.5) and the representation (3.13) of the position vector of the surface in asymptotic coordinates shows that the generatrices are extended as complete straight lines to both sides. In the domain of the parameters (u, v) they are assigned straight lines $u = const$ in the band Π, each of which intersects the graph of the function $v = \dfrac{g'(u)}{f''(u)}$ in only one point (or has no common points altogether with this graph). This means that the entire band Π is mapped by the formulae (3.5) or (3.13) onto one surface S (possibly with self-intersections) consisting of two regular sheets[19] S^+ and S^- corresponding to the domains $D^+ : (u_1, u_2) \times (-\infty < v < \dfrac{g'(u)}{f''(u)})$ and $D^- : (u_1, u_2) \times (\dfrac{g'(u)}{f''(u)} < v <$

---

[19] In reality, the sheet S^- may consist of a finite or even countable set of connected components.

$+\infty$), and on each generatrix $u = c$ there is no more than one singular point – the image of the point $(u = c, v = \dfrac{g'(c)}{f''(c)}) \in \Pi$. On the separating domain D^+ and D^- of the curve $L : g'(u) - vf''(u) = 0$, the following cases are possible.

a) If $f''(u) \equiv 0$, then $f'(u) = c = const$, $f = cu + c_1$, there is no line $L : g'(u) - vf''(u) = 0$ at all, and the domain D^+ coincides with all the band $\Pi : u_1 < u < u_2, -\infty < v < +\infty$, and its image under the mapping (3.7) will be the band P between two parallel straight lines $x + cy = g(u_1)$, $x + cy = g(u_2)$. In this case, all generatrices are parallel to the constant direction $\{-c, 1, c_1\}$, and the surface is a *cylinder* over the band P.

b) Let now be an arc of the line $L : g'(u) - vf''(u) = 0$. Since $g'(u) \neq 0$, then on this line $v \neq 0, f''(u) \neq 0$. We represent it as $v = \dfrac{g'(u)}{f''(u)}$. The curve $L : g' - vf'' = 0$ on the surface S is assigned a line L^* with the position vector $\mathbf{r}^* = \mathbf{a}(u) + \dfrac{g'(u)}{f''(u)}\mathbf{b}(u)$. Consider two possible cases.

b') $\dfrac{g'(u)}{f''(u)} \equiv c = const$. In this case, on the line L we have $g = cf' + c_1$, $y = v = c, x = g - vf' = c_1$, i.e., all the line L is mapped into the point $(x, y) = (c_1, c)$, so the line L on the surface S is assigned one point, in which all generatrices intersect – the surface S is a *cone* consisting of two sheets.

b'') Let the relation $\dfrac{g'(u)}{f''(u)}$ be constant at no interval. The singular line in the parameter region can be presented as a graph of the function $v = v(u) = g'(u)/f''(u)$. First, as in [183], let us assume that $\dfrac{g'(u)}{f''(u)} \in C^1$. Then the singular line on the surface S will be assigned a curve $\Gamma : (x(u), y(u), z(u)$ with some C^1-smooth coordinates of its points, for which

$$x'(u) = v'(u)f'(u), \quad y'(u) = v'(u), \quad z'(u) = v'(u)(f(u) - uf'(u)).$$

$$(3.14)$$

Therefore, for surfaces of class C^3 in points, where $v'(u) \neq 0$, the generatrices on the surface S, as is seen from comparison of the expression (3.9) for their directions with the formulae (3.14), are tangent to the curve Γ. What is more, the set of possible singular points with $v'(u) = 0$ is nowhere dense (in these singular points Γ, as a rule, has cuspidal points, see the book [198]; in our case for isolated singular points with the change of sign $v'(u)$, though, it is readily seen from (3.14)). Thus, between its possible singular points the curve Γ is, by the terminology of [198], *a segment of the envelope* of the family of generatrices of the surface S, thereby representing a *torse*.

The matter is more complicated for surfaces of C^2 smoothness. In this case, the definition of the *discriminant* given in [198] and representing only a *necessary* condition for this locus to be an envelope, is met for the geometric locus Γ of the singular points. In the general case, we can only say right away that Γ is a continuous curve with zero two-dimensional measure[20], which *a priori* may have no tangent at any point, so we would not be able to speak of the surface S as of a *development of tangents* to some curve. But in reality in our case the curve Γ possesses the following unexpected "good" property: at each point $u = u_0$, at which the contingency of the curve L does not contain the horizontal direction $v = const = v(u_0)$, the contingency of the curve Γ at the corresponding point consists of the *only direction coinciding* with the direction of that generatrix on the surface, which arrives to this point of the curve Γ and thereby Γ at these points either has a tangent or possesses one-sided half-tangents having collinear directions (under the direction we understand not a vector but a line of action of the vector). But if at a corresponding point the contingency of the curve L contains the horizontal direction (or even consists of only the horizontal direction), a *direction coinciding* with the direction of the generatrix *will be found* in the contingency of the curve Γ, so that in any case the generatrix can be considered to be somehow "tangent" to the singular line Γ. Indeed, the generatrix arriving to the point of the surface with intrinsic coordinates (u_0, v_0), has the direction of the vector $\mathbf{l} = \{-f'(u_0), 1, f(u_0) - u_0 f'(u_0)\}$, see (3.9). By a sequence of points $(u_n, v_n) \in L \subset \Pi$, in which u_n monotonously tends to u_0, let us have $\dfrac{v_n - v_0}{u_n - u_0} \to a, n \to \infty$, where $0 < |a| < \infty$.

---

[20] To be more exact, its projection to any plane has two-dimensional measure zero.

Then by direct calculation, we verify that $\dfrac{\mathbf{r}(u_n) - \mathbf{r}(u_0)}{u_n - u_0} \to a\mathbf{l}$, i.e., the direction of the generatrix obligatorily enters the contingency of the curve Γ at the point where this generatrix meets with Γ. What is more, we will now show that the contingency of tangents to Γ *always* contains the direction of the generatrix. Let $M_1, M_2, \ldots, M_n, \ldots$ be a (monotonous in the argument u) sequence of points from Γ, which converges to the considered point $M_0 \in \Gamma$, and along which there is a limit position of corresponding secant *rays* $M_0 M_n$. On the curve $L \in \Pi$ this sequence is assigned some sequence of points $\tilde{M}_n(u_n, v_n)$. If in this sequence some subsequence of rays $\tilde{M}_0 \tilde{M}_n$ (where $\tilde{M}_0 \in L$ is the pre-image of the point $M_0 \in \Gamma$) defines the non-vertical and non-horizontal direction of the vector, entering the contingency of the curve L, then by the above proven the corresponding direction on Γ will coincide with the direction of the generatrix. But if all converging subsequences in $\{\tilde{M}_n\}$ define the vertical direction, i.e., $\dfrac{|v_n - v_0|}{|u_n - u_0|} \to \infty, n \to \infty$, then, beginning from some number, we will have $v_n \neq v_0$, and on the curve Γ the directions of the considered secants can be represented as follows. We find the vectors $\mathbf{t}_n = \dfrac{u_n - u_0}{v_n - v_0} \dfrac{\mathbf{r}(u_n) - \mathbf{r}(u_0)}{u_n - u_0}$. By simple transformations, we find that the limit of the vectors $\mathbf{t}_n$ will be the vector $\mathbf{l}$, which is what was required to be proved. If the contingency of the curve L at the point $u = u_0$ contains the horizontal direction and some non-horizontal direction, then this non-horizontal direction in the contingency of the curve Γ is assigned the direction coinciding with the direction of the generatrix at a corresponding point. Finally, if all the contingency of the curve L consists of only the horizontal direction, but by the condition there are no horizontal segments on L, then the sequences of the points $(u_n, v_n) \in L, v_n = max\ v(u), 0 < u \leq u_n$ will be assigned on Γ the rays whose directions converge to the direction of the generatrix at the point u_0, v_0.

Besides, we can obtain at least part of the information on the structure of the surface $S^+ \cup S^-$ in the neighbourhood of points from Γ similar to the case of surfaces of smoothness C^3 and higher. First, from the formula (3.6) we find that both regular sheets S^+ and S^- have on Γ limit positions of their normals with opposite directions.

Second, if in the band Π there exists a smooth curve $\gamma : v = V(u)$ intersecting the singular line $L : v = g'(u)/f''(u)$ at some point $u = u_0$ and such that its image $\gamma_* : x = x_*(u), y = y_*(u), z = z_*(u)$ on S is transversal to the generatrix at the corresponding point, then it will turn out that of necessity $V'(u_0) = 0$ and $x'_*(u_0) = y'_*(u_0) = z'_*(u_0) = 0$, i.e., for any curve on S the transversal to the generatrix at its final point the points of the discriminant are singular. Further, if a plane Π_0 is *found*, which passes through some point $M_0 \in \Gamma$ and intersects in this point the generatrix and both sheets S^+ and S^-, without having in a sufficiently small neighbourhood of the point M_0 other common points with Γ (this will be the case, e.g., if the line $L \subset \Pi$ has an isolated point of transversal intersection with some segment $v = const$), then for the curve $\Gamma_0 = (S \cap \Pi_0)$ the point M_0 will be a cuspidal point of the first type. Indeed, let the point M_0 have the coordinates (x_0, y_0, z_0) with the pre-image at the point $(u_0, v_0) \in \Pi$. Let $\{a, b, c\}$ be the normal to the plane Π_0. As Π_0 intersects the generatrix at the point M_0, then $-af'(u_0) + b + c(f(u_0) - u_0 f'(u_0)) \neq 0$, so in a sufficiently small neighbourhood (u_0, v_0) the curve γ_0 – the pre-image of the section of the surface S by the plane Π_0 – has the equation $v = V(u) \in C^1$. Calculating $V'(u_0)$ from this equation or, simpler, from the relation

$$a(x(u, V(u)) - x_0) + b(V(u) + c(z(u, V(u)) - z_0) = 0, \qquad (3.15)$$

we obtain $V'(u_0) = 0$. Therefore, for the curve

$$\Gamma_0 : (x(u, V(u)), y(u, V(u)), z(u, V(u))) \subset S$$

we have $x'(u_0) = 0, y'(u_0) = 0, z'(u_0) = 0$; what is more, by straightforward calculations, it is verified that the limit directions of the tangents on the right-hand and left-hand side coincide in the singular point. Besides, the branches of the curve Γ_0 in the neighbourhood of the point M_0 are situated on *different sides* of their common semi-tangent in M_0. A long but elementary verification of this fact is as follows. Draw through the semi-tangent to Γ_0 at the point M_0 a plane Π_1 perpendicular to the plane Π_0. Then substitute into the left-hand side of the equation of the plane Π_1 the equation of the curve Γ_0 and, taking into account the fact that this curve lies on the plane Π_0, obtain the expression $F(u) = (a^2 + b^2 + c^2)[v(f_0 - f(u)) + (u_0 - u)(g(u) -$

$vf'(u)) + G(u) - G(u_0)]$, where $f_0 = f(u_0)$, and $v = v(u)$ is taken from the equality (3.15). We have $F(u_0) = 0$, and the derivative $F'(u)$ proves to be of the form $(1 + o(u - u_0))(u - u_0)(g'(u) - vf''(u))$, i.e., is of constant sign, so $F(u)$ in transition through the point $u = u_0$ changes the sign, therefore, in the neighbourhood of the point M_0 the curve Γ_0 lies on *different* sides of its semi-tangent, having in M_0 a first-type cuspidal point. This means that the surface cannot be extended over the line Γ not only as a smoothly developable but even as a simply developable piecewise smooth surface.

With respect to the possibility of the existence of a piecewise smooth developable surface, see [163].

(5) Consider now the cases when different types of surfaces – a torse and a cylinder or a torse and a cone – occur at some point $u = u^*$. In the first case, let the (left from point $u = u^*$) part of the band Π be assigned a cylindrical surface, and there be a torse on the right, i.e., this is the development of the tangents. Then at the point u_* we have $f''(u^*) = 0$ and the line L when approaching u^* from the right should go into infinity, having the vertical asymptote $u = u^*$, which is assigned the common generatrix of the cylindrical part and torsial part of the surface. When we study the smoothness of the surface, we consider only the neighbourhood of any fixed point, in this case the neighbourhood of the point (u^*, v) with the finite value v such that $(u^*, v) \notin L$, so that in a sufficiently small neighbourhood of this point we conclude from the formulae (3.10) that for the values of the second derivatives of z when approaching (u^*, v) both from the left and right sides of the band Π we have the equal finite limits $\dfrac{1}{g'(u^*)}$ for z_{xx} etc. A similar reasoning is also valid for the "glueing" of a conical part and a torsial part of a surface; it should only be noted that the vertex of the cone will be positioned on the cuspidal edge of the torsial part of the surface.

(6) It remains for us to show that the C^2-smooth developable surface without points of flattening cannot consist of: (a) a cylindrical part and a conical part; (b) a cylindrical part and a torse with a finite length of the common generatrix; (c) parts of two cones with different vertices; (d) parts of two torses without intersecting discriminants[21].

---

[21] A similar problem was considered in [163] from a slightly different viewpoint.

Indeed, if a surface is glued from parts of a different nature, then due to the unicity of the generatrix passing through each point, these parts should be glued along their common generatrix. Let in the representation (3.5) of a C^2-smooth surface S the segment $[u_1, u_0]$, correspond to the cylindrical part; and the segment $[u_0, u_2]$, to the conical or torsial part. Then at the point $u = u_0$ we have $f''(u_0) = 0$, and the satisfaction of the relation $g' - vf'' = 0$ on the right-hand side leads to the equality $g'(u_0) = 0$ (if $v_0 \neq \infty$), which contradicts the condition $g'(u) \neq 0$. But if two conical parts with different poles, or a torse and a cone with its vertex not on the cuspidal edge, or two torses with boundary generatrices emanating from different points are glued, then we obtain that there are two singular points on the generatrix common for two parts of the generatrix, which is impossible.

But note that a cylinder and a torse or two torses can be glued together, if the cuspidal edge at the torse(s) goes to infinity having the common generatrix as an asymptotic line to the cuspidal edge.

We complete the proof and discussion of the theorem with the following remarks.

Remark 3.5. In the case of smoothness of class C^2 it may prove that on the surface S there is no plane (or none at all) piecewise smooth curve with a smooth pre-image $v = V(u)$ in the band Π, which would intersect the singular manifold Γ at some isolated point M_0 transversally to the generatrix (it is borne in mind that the arc of the curve at each sheet $S^{\pm}$ in the sufficiently small neighbourhood of the point M_0 is smooth with the probable disturbance of smoothness only at the point M_0). This is proved by the existence of an example of nowhere differentiable continuous function $v = v_0(u) = g'(u)/f''(u) = \sum 2^{-n} \sin(8^n u)$, for which at each point the contingency of tangents contains the half-plane; therefore, no smooth curve $v = V(u)$ can intersect the curve $v = v_0(u)$ at an isolated point (the existence of this function was kindly reported to me by T.P. Lukashenko).$\bullet$

Remark 3.6. In part **(2)** of the proof we gave the formula (3.11) of asymptotic parametrization, for which it was asserted (with reference to [80] at $n = 2$) that the position vector $\mathbf{a}(u)$ of the directrix of a surface of class C^n belongs to class C^{n-1}. In reality, there is a reparametrization in which $\mathbf{a} \in C^n$ (but $\mathbf{b}$ will as previously be in class C^{n-1}).●

Indeed, for this in the formulae (3.5) it is sufficient to pass from the parameter u to the parameter $x = g(u)$. Then the directing curve with $v = 0$ will be represented as $\mathbf{a}(x) = \{(x, 0, z = xu(x) - \int_{x_0}^{u(x)} g(t)dt)\}$, and $\mathbf{b}(x) = \{-f'(u(x)), 1, f(u(x)) - u(x)f'(u(x))\}$, and as $z'_x = u(x) = g^{-1}(x) \in C^{n-1}$, then for $\mathbf{a}(x)$ and $\mathbf{b}(x)$ we will obtain the proposed smoothnesses.

Remark 3.7. From the formula (3.10) it is seen that on the singular line Γ the derivative z_{xx} becomes infinite, so even in an analytical case none of the closed sheets S^+ and S^- (i.e., with inclusion of the edge) is not of class C^2, though the open surface and its edge as a curve separately can well be analytical.●

Remark 3.8. Note that the formula (3.5) yields for the equation (3.4) the solution of the Cauchy problem with the initial conditions:

$$z'_x(x, 0) = \varphi(x)(\varphi'_x \neq 0), \, z'_y(x, 0) = F(x),$$

which could be transformed as follows: assuming $u = \varphi(x), x = g(u)$, we obtain $z(x, 0) = ug(u) - \int g du, \, z'_y(x, 0) = f(u) = F[g(u)]$.

Remark 3.9. The Monge–Ampère equations, more general than (3.4) but also parabolic in the sense that their solutions describe developable surfaces, have important applications in mathematical physics, in particular, in meteorology. In this respect, the reader may refer to the papers [118–121, 138–141] and other works by the same authors. Some applications of ruled surfaces (irrespectively of the Monge–Ampère equations) can be seen in [9]. Important applications of such developable surfaces as a cylinder and a cone in the theory of thin elastic shells can be found in [125].●

3.2.2   *Additional properties of developable surfaces*

From the proof of Theorem 3.6, we have several useful corollaries verifying the structure of developable surfaces.

Corollary 3.2. *If on the generatrix of a C^2-smooth surface of zero curvature there is a point of flattening, then the entire generatrix consists of points of flattening.*

Indeed, let on the generatrix there be points not being points of flattening. Then by Theorem 3.6 within a sufficiently narrow strip along the generatrix the surface is representable as (3.5), and there is the formula (3.10), from which it is seen that there are no points of flattening along the entire generatrix either.

Thus, the proposition of this corollary presents a sharpening for C^2-smooth surfaces with an l. E. metric of the property of C^1-smooth developable surfaces on that either all points of a generatrix are planar points or there are no planar points on it.

Corollary 3.3. *For a developable surface without points of flattening to be of smoothness $C^n, n \geq 2$, it is necessary and sufficient that it has asymptotic parametrization*

$$\mathbf{r}(u, v) = \mathbf{a}(u) + v\mathbf{b}(u), \ \ \mathbf{b}(u) \not\parallel \mathbf{a}'(u), \ (\mathbf{a}'\mathbf{b}\mathbf{b}') = 0, \qquad (3.16)$$

with the directrix $\mathbf{a}(u) \in C^n$ and the directions of the generatrices $\mathbf{b}(u) \in C^{n-1}$; what is more, in the general case, it cannot be asserted that for a surface of class C^n there is asymptotic parametrization with $\mathbf{b}(u) \in C^n$.

This proposition at $n = 2$ has been proved in [80]. For general $n > 2$ it follows from parts (1), (2), (3) of the proof of Theorem 3.6 and Remark 3.6.

The latter corollary admits the following refinements.

If we eliminate the condition of the absence of points of flattening, then only a much lesser smoothness can be guaranteed for asymptotic parametrization.

Proposition 3.2. *For any developable surface of smoothness C^n, $n \geq 1$, each point has a neighbourhood, in which there is an asymptotic parametrization of the form (3.11) with $\mathbf{a}(u) \in C^n$ and $\mathbf{b}(u) \in C$.*

The proof of this proposition, given in [76] for $n = 2$, is rather lengthy. Let us present a shorter proof. If the point considered is planar, the proof is evident. Let the neighbourhood of a nonplanar point M_0 on a surface S be considered, and let the coordinate system be chosen such that M_0 is the origin, the generatrix passing through M_0 goes along the axis y, and the tangent plane coincides with the plane xy. We draw the section of the surface by the plane xz and obtain some curve $\gamma : (x, 0, z(x, 0))$ of smoothness C^n. The set of planar points on this curve is either empty or consists of no more than a countable set of open arcs, and nonplanar points cannot be isolated. We draw through all nonplanar points the unique generatrices passing through them. We assume that the surface S is considered over the square $Q : |x| \leq \varepsilon, |y| \leq \varepsilon$. No generatrix can be situated on the plane xz, as otherwise two points of the surface $z = z(x, y)$ will prove to be in Q over the axis Ox. Therefore, all generatrices will have directions of the form $\{l_1(x), 1, l(x)\}$, where $l(x) = l_1(x)z_x(x, 0) + z_y(x, 0)$. As the generatrices go from edge to edge of the surface, their projections go from edge to edge of the square Q; further, on the surface with the unique projection to the surface xy there is no intersection of not only the generatrices on the surface, but also of their projections on the plane xy (otherwise we would have had two points of the surface $z = z(x, y)$ over the point of intersection of the projection of the generatrices). Hence, it follows that the inequality $|l_1(x)| < |x|/\varepsilon$ should be satisfied for the component $l_1(x)$, i.e., at $x \to 0$ we have $l_1(x) \to 0 = l_1(0)$. Thus, if the nonplanar point M_0 belongs to some interval $(-\delta, \delta) \subset (-\varepsilon, \varepsilon)$, over which the arc γ consists of nonplanar points, then the continuity of the family of generatrices in M_0 is proved. Let now be no such arc. On the curve γ consider an arc γ_0 over some interval $(a, b) \subset (-\varepsilon, 0) \cup (0, \varepsilon)$, consisting of planar points. Then on the surface S it is assigned a plane strip between two (not necessarily parallel) generatrices passing through the endpoints of the arc (here we have to take into account the structure of plane domains on a developable surface, described in Subsection 3.1.2). In this case, this band can be spanned by a continuous family of non-intersecting generatrices, also continuously contiguous to the extreme generatrices of the band. As the result, we have that the component $l_1(x)$ is continuous at the point $x = 0$, and after applying similar arguments to the other points $x \in (-\varepsilon, \varepsilon)$ we obtain a continuity within the entire interval $(-\varepsilon, \varepsilon)$. As the result, the C^n-smooth curve γ, as a directrix, and the vectors $b(x) = \{l_1(x), 1, l(x)\}$, as the directions of the generatrices, yield the sought-for asymptotic parametrization of the surface.

Remark 3.10. A more exact proposition on the character of regularity of generatrices' directions is that in the representation (3.11) the vector $\mathbf{b}(u)$ can always be chosen to be of Lipschitz class $C^{0,1}$, for which $\mathbf{a}(u) \in C^n, \mathbf{b}(u) \in C^{0,1}$. This gives an unexpectedly simple answer to the question 3.4 of the description of C^1-smooth torses. For details, see [149].

Indeed, in the construction and notation of the proof of Proposition 3.2 from the condition of non-intersectionability of the projections of the generatrices inside the square Q we obtain that the values of the component $l_1(x)$ at the points x_1, x_2 with $|x_1| < |x_2| \le \varepsilon/2$ should satisfy the inequality

$$|l_1(x_2) - l_1(x_1)| \le \frac{2}{\varepsilon}|x_2 - x_1|.$$

This means that the choice of the directrix and its parametrization, made in the course of the proof of Proposition 3.2, yields the asymptotic parametrization (3.11) of the surface, for which $\mathbf{a}(u) \in C^n, \mathbf{b}(u) \in C^{0,1}$. $\bullet$

Further, it is not difficult to show the validity of the following:

Proposition 3.3. *If in a neighbourhood of a point $M \in S$ the surface S is of class $C^n, n \ge 2$, and has some asymptotic parametrization of the form (3.11) with $\mathbf{a}(u) \in C^n$ and $\mathbf{b}(u) \in C^{k,\alpha}, 1 \le k + \alpha \le n$, in which $\mathbf{a}' \not\parallel \mathbf{b}$, then for any directrix of smoothness C^n, passing through the point M and not tangent to the direction of the generatrix in it, the surface has asymptotic parametrization of the same smoothness as initial parametrization.*

For the proof, it is sufficient to represent the position vector $\mathbf{A}(s)$ of the new directrix as $\mathbf{A}(s) = \mathbf{a}(u) + v(u)\mathbf{b}(u)$ with $s = s(u)$ and from the equality $\mathbf{r} = \mathbf{a}(u) + v\mathbf{b}(u) = \mathbf{A}(s) + w\mathbf{b}(s)$, where $\mathbf{b}(s(u)) = \mathbf{b}(u)$, we can obtain the required proposition.

The next proposition shows that the choice of asymptotic parametrization with bringing the regularity $\mathbf{b}(u)$ up to class C^1 is not always possible.

Proposition 3.4. *If there is at least one generatrix with points of flattening on a surface, then even in the class of smoothness C^∞*

I.Kh. SABITOV

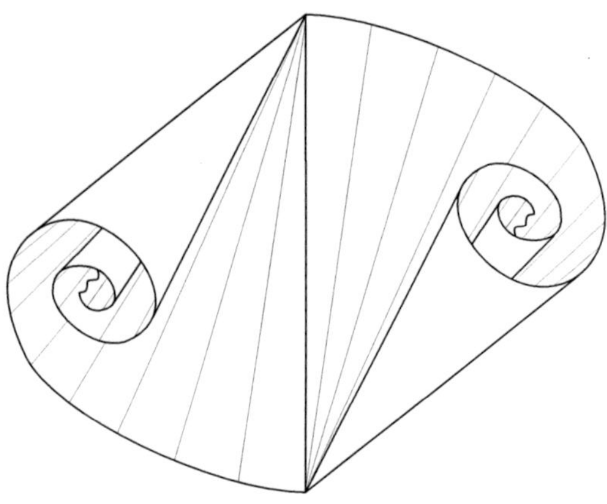

Figure 2

it is possible to indicate surfaces, for which there is no asymptotic parametrization (3.16) with $\mathbf{b}(u) \in C^1$.

An example of such a surface is given in [102] (pp. 68–69) and examined in detail in [180]. The structure of this surface is seen in Figure 2 to be composed of two conical parts with different vertices, and, although there is no explicit equation of the surface, it is perceived from explicitly written-out C^∞-smooth first and second quadratic forms to belong to class C^∞. What is more, the first form has the standard Euclidean form $du^2 + dv^2$, and both forms jointly satisfy the Peterson–Codazzi equations. If we will try to represent the equation of this surface as an equation of a ruled surface, then the directrix will be of smoothness C^∞, whereas the family of vectors of the generatrices will prove not C^1-smooth.

According to [180], the absence of a negative effect of the discontinuity of the first derivative of the generatrix direction field on the general smoothness C^∞ of the surface is explained by the fact that the derivative is discontinued for objects located on the tangent plane: the vectors $\mathbf{b}(u)$, their derivatives $\mathbf{b}'(u)$ and the derivative $\mathbf{a}'(u)$ remain on the tangent plane during the change of u, not disturbing its behaviour (it could also be said that the vector product of the "bad" vectors $\mathbf{b}$ and $\mathbf{b}'$ can yield a "good" normal determining the "good" behaviour of the surface). An example of such a situation can be easily modelled by constructing a nonsmooth family of straight lines on a usual plane – the family is nonsmooth, but the plane remains an analytical surface.

The same work [180] shows how to choose the "buffer" zone in the band near the common generatrix (with points of flattening) between surfaces of different shapes, so that they yield a regular surface in glueing, but so that herewith there be no regularity in asymptotic parametrization. If the representation (3.5) is used for this purpose, then the issue is reduced to finding C^∞ functions $g(u)$ and $f(u)$ given in the intervals $[u_1, u_0)$ and $(u_0, u_2]$, for which the functions z_{xx}, z_{xy} and z_{yy} from (3.8) are equal to zero on the generatrix $u = u_0$ and have a smoothness of class C^∞ on it and in its neighbourhood, in the assumption that $z_{xx} \neq 0$ is outside the generatrix $u = u_0$.

Further, in [76] it is asserted and an example is given that in an analytical case it is possible that any asymptotic parametrization will be "bad" with $\mathbf{b}(u) \notin C^1$ too. However, [180] indicates an error in the consideration of this example and proves the following:

Proposition 3.5. *Analytical asymptotic parametrization always exists in the analytical case in a sufficiently small neighbourhood of any point.*

We adduce this proof in a broader exposition than in [180]. Let M_0 be an arbitrary point of a developable analytical surface S. As there can be no planar points on a nonplanar analytical surface, M_0 is a nonplanar point. Choose the coordinate axes such that the generatrix passing through M_0 coincide with the axis y, and the tangent plane to the surface be the plane xy. The plane xz intersects the surface by the analytical curve $\gamma : x = x, z = z(x, 0)$ containing not a single rectilinear segment. Let the generatrix passing through the point $(x, 0, z(x, 0))$ have the direction $\{l_1(x), 1, l(x)\}$, where $l(x) = l_1(x)p(x, 0) + q(x, 0)$. We have $l_1(0) = 0$ and $l(0) = 0$. From the proof of Proposition 3.2 we know that the family of the generatrices is continuous. As along the generatrices the functions $z_x(x, y)$ and $z_y(x, y)$ are constant, we have

$$z_x(x + tl_1(x), t) = z_x(x, 0), \quad z_y(x + tl_1(x), t) = z_y(x, 0), \quad \forall t \in (-\varepsilon, \varepsilon).$$

Differentiating the first equality with respect to t at point $t = 0$, we obtain the equation $z_{xx}(x,0)l_1(x) + z_{xy}(x,0) = 0$. If $z_{xx}(0,0) \neq 0$, then we have that the function $l_1(x)$, and together with it $l(x)$, are analytical. But if $z_{xx}(0,0) = 0$, then $z_{xy}(0,0) = 0$, too; expanding them in a power series of x, we obtain that the relation $z_{xy}(x,0)$: $z_{xx}(x,0) = -l_1(x)$ is analytical (as it is already known that this relation is continuous and $l_1(0) = 0$, this means that the integer order of zero in the function $z_{xy}(x,0)$ is greater than in $z_{xx}(x,0)$, which was sufficient to show, as it yields the analyticity of the asymptotic parametrization $\mathbf{r}(x,t) = \{x, 0, z(x,0)\} + t\{l_1(x), 1, l(x)\}$.

Proposition 3.5 can be refined as follows.

Proposition 3.6. *If, on an analytical developable surface S, we single out some analytical directrix L, which is nowhere tangent to the generatrices and is situated in the interior of S, then, based on this directrix, there exists an asymptotic parametrization of the surface along the entire curve L with an analytical dependence of the direction of the generatrices on the analytical parametrization of the curve.*

Proof. First, let us perform constructions almost as in the proof of the previous proposition. Let M_0 be some point on an analytical developable surface $S : z = z(x,y)$, and L be an analytical curve on S, passing through the point M_0. We assume that the coordinate axes are chosen such that M_0 has the coordinates $(0,0,0)$ and the plane xy is tangent to S at M_0, while the tangent to L at the point M_0 goes along the axis x. Let us draw through the point M_0 the section of the surface S by the plane xz. Let $\gamma : (u, 0, z(u,0))$ be a curve obtained in the section. As the generatrix is orthogonal to the vector $\{0,0,1\}$ and is not tangent to L, it will not be orthogonal to the axis y, but then as in the proof of the previous proposition we can show that the asymptotic parametrization $\mathbf{r}(u,t) = \{u, 0, z(u,0)\} + t\{l_1(u), 1, l(u)\}$ based on the directrix γ is analytical. Let

$$x = x(v), y = y(v), z = z(v) = z(x(v), y(v))$$

be some analytical parametrization of the given analytical curve L with $y'(0) = z'(0) = 0, x'(0) \neq 0$ (it is assumed that the point $M_0 \in L$ is assigned the value of the parameter $v = 0$). Let $\mathbf{r}(v, \tau) = \{x(v), y(v), z(v)\} + \tau\{a(v), b(v), c(v)\}$ be an asymptotic parametrization based on the directrix L, where $\mathbf{l}(v) = \{a, b, c\}$ is the unit vector of the generatrix at the corresponding point of the directrix. The generatrix passing through the point $(x(v), y(v), z(v))$ on the curve transverses the curve γ at some point $(u, 0, z(u, 0))$, so this generatrix has the direction $l_1(u) = \{s_1(u), 1, s_(u)\}$. Therefore, at the point of its intersection with γ we have an equality $F(u, v) \equiv u - x(v) - y(v)s_1(u) = 0$. As $y(0) = 0$, then $F_u(0, 0) \neq 0$, which means that from this equality u is expressed as an analytical function of $v : u = u(v)$. Then it will turn out that the vector $l(v)$ coincides with the unit vector of the direction $l_1(u(v))$, so its dependence on v is analytical, which is sufficient for the proved proposition to be valid, as transitions in analytical local maps along L are analytical (except, possibly, the first and last maps in the case of the closed curve L).

Sometimes, we have to search for developable surfaces with a directrix in the form of a preset curve, which, as a line on the surface, should possess some special properties. Consider, e.g., the case when the directing curve $\Gamma : \mathbf{r} = \mathbf{r}(u)$ on the sought-for surface $S \in C^2$ should be *geodesic*. Let the curve $\Gamma \in C^2$ and let its curvature $k(u)$ be distinct from zero. Let us investigate which additional conditions this curve should satisfy to be a geodesic on a developable surface $S : \mathbf{r}(u) + v\mathbf{l}(u) \in C^2$ (being of class C^2 is assumed in the general sense, but not necessarily in the coordinates (u, v)). As the curvature of the geodesic directrix is distinct from zero, the surface S contains no points of flattening, so, by Corollary 3.3 and Proposition 3.3, its generatrices $\mathbf{l}(u) \in C^1$. Besides, as in the geodesic line the principal normal $\mathbf{N}$ is parallel to the normal $\mathbf{n}$ of the surface, we can consider that $\mathbf{N} = \mathbf{n} \in C^1$. We have that one additional condition for the curve $\Gamma \in C^2$ is that its principal normal $\mathbf{N}$ should be of class C'^1. This means that there exists $\mathbf{N}' = -k\mathbf{T} + \varkappa\mathbf{B}, \mathbf{B} = [\mathbf{T} \times \mathbf{N}] \in C^1$, where $\mathbf{T}$ is a unit tangent vector, and the coefficient k is determined by the equality $\mathbf{T}' = k\mathbf{N}$. We have $\exists \mathbf{B}' = -\varkappa\mathbf{N}$, i.e., there should be a torsion of the curve Γ. As $\mathbf{l} \perp \mathbf{n} = \mathbf{N}$, we will search for $\mathbf{l}(u)$ in the form

$$\mathbf{l}(u) = \cos\alpha(u)\mathbf{T}(u) + \sin\alpha(u)\mathbf{B}(u), \qquad (3.17)$$

where the angle $\alpha(u)$ is determined from the equality

$$k(u)\cos\alpha(u) + \varkappa(u)\sin\alpha(u) = 0,$$

obtained from the conditions $(\mathbf{T}, \mathbf{l}, \mathbf{l}') = 0$. The existence of this angle α with the inequality $\sin \alpha \neq 0$ satisfied is provided for by the condition $k(u) \neq 0$. As the normal to the surface S in this choice of $\mathbf{l}(u)$ is collinear with the principal normal, then the curve Γ proves to be a geodesic; what is more, the standard Euclidean coordinates (x, y) can be chosen such that the line Γ corresponds to the segment $y = 0$. Additionally, we obtain that the relation $\dfrac{\varkappa}{k}$ should be of smoothness C^1. Such a curve is, e.g., a plane C^2-smooth curve, though even not any C^3-smooth spatial curve satisfies all these conditions, but the curves of smoothness C^4 obviously do. This proves the following proposition:

Proposition 3.7. *Any curve* $\Gamma : \mathbf{r} = \mathbf{r}(u) \in C^4$ *of curvature not equal to zero can locally be considered as a geodesic directrix at some (unique) developable surface of class* C^2, *which is an isometric embedding of the standard metric* $dx^2 + dy^2$ *into* R^3 *with a mapping of a segment of the axis* Ox *into the curve* Γ.

For the case of a geodesic $\Gamma \in C^2$, in the end, we obtain that it has a C^1-smooth Frenet trihedron with continuous torsion $\varkappa$; moreover, the relation $\varkappa/k$ should be a C^1-smooth function. Similarly, considering the general case of smoothness of class $C^n, n \geq 2$, we obtain that the geodesic lines on a developable surface $S \in C^n$ either have a smoothness of no less than C^{n+2} or are examples of curves of smoothness C^n having a torsion $\varkappa$ of smoothness C^{n-2}, as well as curvature k; what is more, the relation $\varkappa/k$ should be of smoothness C^{n-1}. Such curves, for which the smoothness of torsion (and, in general, curvatures of the order m, if we are talking of curves in a multidimensional space E^p, $p \geq m + 1$, considering the usual curvature as the curvature of order 1) is obviously greater than its formally calculable smoothness, have not been studied sufficiently fully.●

Further, from (3.17) we can see the validity of the following:

Proposition 3.8. *If on a* C^2-*smooth developable surface there is a* $C^n(n > 3)$-*smooth (analytical) geodesic directrix of curvature distinct from zero, the surface is in reality smooth of class* C^{n-2} *(analytical).*

If the curvature of the given geodesic Γ goes to zero, then there is one evident case with curvature identically equal to zero, and the second case, when the curvature Γ has isolated integer-order zeros. Let us require more, namely, let us assume that Γ is an analytical curve and prove the following proposition:

Proposition 3.9. *If on a C^2-smooth developable surface there exists an analytical geodesic directrix, then the surface is in reality analytical and its generatrices have directions* $\mathbf{l}$ *determined by the formula (3.17).*

Let us prove this proposition. As at present the condition $k \neq 0$ is absent, we cannot assert *a priori* that the directions $\mathbf{l}$ of the generatrices are of class C^1, as now points of flattening can be on the surface. Therefore, we will do as follows.

Let us first introduce an analytical curvature k_A of an analytical curve Γ. At points of the curve Γ, in which its curvature is not equal to zero, it is usually assumed that $\mathbf{r}_{ss} = k\mathbf{N}$, where $\mathbf{r}(s)$ is the position vector of the curve; $\mathbf{N}$, its principal normal; and $k = |\mathbf{r}_{ss}| > 0$, the curvature of the curve. But in transition through the points $\mathbf{r}_{ss} = 0$ the direction of the principal normal should be chosen such that it changes continuously. In our case the line Γ is geodesic on the surface, so its principal normal $\mathbf{N}$ is collinear with the normal $\mathbf{n}$ to the surface. Therefore, along Γ we already have a continuous vector field $\mathbf{n}$, which at the initial region, where $k \neq 0$, can be taken to coincide with $\mathbf{N}$ (due to the freedom of choosing the direction of the normal to the surface at the initial point), and after the transition through the possible zero of curvature we will leave the direction of the principal normal $\mathbf{N}$ coincident with $\mathbf{n}$, i.e., within the entire range of Γ we assume $\mathbf{N} = \mathbf{n}$. Then we will, respectively, set $\mathbf{r}_{ss} = k_A\mathbf{n}$, where the function $k_A = (\mathbf{r}_{ss} \cdot \mathbf{n})$ plays for the curve Γ the role of curvature with a sign, with the behaviour of the sign in transition of $|\mathbf{r}_{ss}|$ through zero at some point $s = s_0$ by the following rule: if $|\mathbf{r}_{ss}|^2 = c_0^2(s - s_0)^{2m}(1 + \ldots)$, $c_0 > 0$, then at even m the function k_A goes via zero without changing the sign, and at odd m it changes its sign. Formally, it looks like as follows: if the sign of k_A for values of $s < s_0$ sufficiently close to s_0 is designated as $sign\, k$, then in the neighbourhood of the point $s = s_0$ we take $k_A(s) = (sign\, k)c_0(s - s_0)^m(1 + \ldots) = (sign\, k)|\mathbf{r}_{ss}|$ at m even and $k_A(s) = -(sign\, k)c_0(s - s_0)^m(1 + \ldots) = (sign\, k)(sign\,(s_0 - s))|\mathbf{r}_{ss}|$ at m odd.

Then for the principal normal $\mathbf{N}$ we have an expression $\mathbf{N} = \dfrac{1}{|\mathbf{f}|}\mathbf{f}$, where the analytical vector $\mathbf{f}$ is determined by the formula: $\mathbf{r}_{ss} = c_0(s - s_0)^m \mathbf{f}(s)$, $|\mathbf{f}(s_0)| \neq 0$. Therefore, both the curvature k_A and the principal normal $\mathbf{N}$ are analytical.

After the introduction of an analytical vector of the principal normal $\mathbf{N} = \mathbf{n}$ on the entire curve, let us now determine the analytical binormal $\mathbf{B}$ on it by the formula $\mathbf{B} = [\mathbf{r}', \mathbf{n}]$. The derivative $\mathbf{B}'$ is collinear to $\mathbf{n}$, so let us set $\mathbf{B}'(s) = -\varkappa(s)\mathbf{n}$, by definition taking the analytical function $\varkappa(s) = -(\mathbf{B}' \cdot \mathbf{n}) = (\mathbf{n}, \mathbf{n}', \mathbf{r}')$ for the torsion[22]. We have

$$\mathbf{B} = [\mathbf{r}' \times \mathbf{N}], \ \ \mathbf{r}' = [\mathbf{N}, \mathbf{B}], \ \ \mathbf{N} = [\mathbf{B} \times \mathbf{r}'],$$

from where by direct verification we obtain that Frenet relations are valid for the introduced trihedron $\mathbf{T} = \mathbf{r}', \mathbf{N}, \mathbf{B}$ and the introduced functions of curvature k_A and torsion $\varkappa$. But it turns out that when a curve is a geodesic and a directrix of a developable surface, then its torsion possesses some special property, namely, *the torsion of necessity becomes zero at points, where the curvature $k_A = 0$; what is more, at these points it will have zero of no smaller order than the curvature.*

Let us prove this property. If we knew that the directions of the generatrices are of class C^1, then we would have made use of the formula (3.17) and would have immediately obtained the required proposition. But we do not know this in advance, so we will do otherwise.

We will regard our surface as an envelope of the family of its tangent planes $\mathbf{n}(s)\mathbf{R} - \mathbf{n}(s)\mathbf{r}(s) = 0$, where $\mathbf{R} = (x, y, z)$; $\mathbf{r}(s)$ is the position vector of the geodesic directrix Γ; and $\mathbf{n}$ is the normal to the surface, coinciding with the principal normal $\mathbf{N}$ of the curve Γ. Then the generatrices will be straight lines in the intersection of two planes

$$\mathbf{N}(s)\mathbf{R} - \mathbf{N}(s)\mathbf{r}(s) = 0 \ \ and \ \ \mathbf{N}'(s)\mathbf{R} - \mathbf{N}'(s)\mathbf{r}(s) = 0.$$

---

[22] From the classical formula for torsion, it is easy to deduce that generally in the analytical case in a point where the curvature turns to zero, this formula has the finite limit, so the definition of the torsion of any analytical curve can be extended in zeros of curvature as an analytical function.

Searching for the position vectors of these straight lines in the form $(x, y, z) = \mathbf{l}(s)v + \mathbf{r}(s) = (a\mathbf{N} + b\mathbf{T} + c\mathbf{B})v + \mathbf{r}(s)$ with $a^2 + b^2 + c^2 = 1$, we will find that at points, where $k_A \neq 0$, we have $a = 0$, $k_A(s)b(s) = \varkappa(s)c(s)$. Therefore,

$$\mathbf{l}(s) = \cos\alpha\mathbf{T} + \sin\alpha\mathbf{B} = b\mathbf{T} + c\mathbf{B},$$

i.e., $k_A \cos\alpha = \varkappa \sin\alpha$. As $\sin\alpha \neq 0$, we can write $\varkappa(s) = \cot\alpha k_A$, whence follows the proved proposition. Besides, we have shown that the directions of the generatrices are analytical functions of the arc-length of the line Γ, whereby the analyticity of the surface S was proved.

Now we can even write down how the position vector of this surface is expressed via curvature and torsion of its directrix geodesic. We have $\sin\alpha \neq 0$, so we can assume $\sin\alpha > 0$. Then we can write down that

$$\sin\alpha = \frac{1}{\sqrt{1 + h^2(s)}}, \quad \cos\alpha = \frac{h(s)}{\sqrt{1 + h^2(s)}}, \qquad (3.18)$$

where the function

$$h(s) = \frac{\varkappa(s)}{k_A(s)}$$

is analytical if its definition is considered to be extended in zeros of curvature with respect to continuity. This means that the position vector of the surface S has the form

$$\mathbf{r}(s) + v\left(\frac{h(s)}{\sqrt{1 + h^2(s)}}\mathbf{T} + \frac{1}{\sqrt{1 + h^2(s)}}\mathbf{B}\right). \qquad (3.19)$$

4   Isometric immersions of locally Euclidean metrics into E^3

4.1   Main results

Thus, in the previous Chapter we studied in sufficient detail the form of the intrinsic geometry of surfaces carrying l. E. metrics. In this Section, we investigate the possibility of isometric embedding and immersion into E^3 of (simply or multiply connected) plane domains with l. E. metrics. The main result is that if a domain with an l. E. metric admits an isometric immersion into E^2, then it is isometrically embeddable into E^3 (which means, in particular, that each simply connected domain with an l. E. metric isometrically embeds into E^3). Further, if a domain is doubly connected, it always isometrically immerses into E^3, and for triply connected domains with l. E. metrics there are already examples of domains unimmersible into E^3.

Now we pass to exact propositions. Let an l. E. metric ds^2 be given in a circular n-connected domain $\bar{D}_n$ (i.e., in a circle – at $n = 1$ and in a domain bounded by n circumferences – at $n \geq 2$) of a plane of variables (ξ, η). Without loss of generality, we can consider that the coordinates (ξ, η) are isothermal (the general case is reduced to this by some quasiconformal transformation, see [184]). Then

$$ds^2 = \Lambda^2(\xi, \eta)(d\xi^2 + d\eta^2), \ \Lambda > 0, \Delta \ln \Lambda = 0. \tag{4.1}$$

Assume further that the regularity $\Lambda(\xi, \eta)$ in the closed domain $\bar{D}_n$ is such that the unique (up to the motion in E^2) mapping $z = x + iy = \Phi(\zeta)$, $\zeta = \xi + i\eta$, reducing (4.1) to the standard form $ds^2 = dx^2 + dy^2$, has in $\bar{D}_n$ the smoothness of class C^m, $m \geq 2$. Let us recall that the mapping $\Phi(\zeta)$ is determined via $\Lambda(\zeta)$ by the equality $|\Phi'(\zeta)| = \Lambda(\zeta)$; therefore, if $\Lambda \in C^{m,\alpha}, m \geq 0, 0 < \alpha < 1$, then this is sufficient for the mapping $\Phi(\zeta)$ to have the smoothness $C^{m+1,\alpha}$.

Theorem 4.1. *Let a mapping $z = \Phi(\zeta)$ of smoothness $C^m(\bar{D}_n)$, $m \geq 2$, be single-valued in $\bar{D}_n$ (i.e., $\bar{D}_n$ with the metric (4.1) is isometrically embeddable into E^2). Then the domain $\bar{D}_n$ with the metric (4.1) admits an isometric embedding into E^3 as a developable surface of smoothness C^∞ in D_n and of smoothness C^m in $\bar{D}_n$. (See [143].)*

Corollary 4.1. *The circle $\bar{D}_1$ with an l. E. metric (4.1) admits an isometric embedding into E^3 as a developable surface of corresponding smoothness.*

Theorem 4.2. *Let in a domain $\bar{D}_2$ (i.e., in a concentric ring) with an l. E. metric (4.1) the mapping $z = \Phi(\zeta)$ be nonsingle-valued. Then $\bar{D}_2$ can be isometrically immersed into E^3 as a developable surface of any smoothness $C^m(D_2), 2 \leq m \leq \infty$. (See [143].)*

This theorem solves negatively the problem raised in [66] (Subsection 3.2.1 (F)(c'')) in that a flat metric not admitting an isometric immersion into E^3 can exist on the cylinder $S^1 \times [0, 1]$. As for the case of l. E. metrics given in domains of connection $n \geq 3$, here already for triply connected domains we show examples of metrics not immersible into E^3 even as a C^1-smooth torse, in particular, as a C^1-smooth surface of extrinsic curvature bounded in the Pogorelov sense[23].

4.2   Proofs of theorems

4.2.1   Proof of Theorem 4.1

We begin with the following lemma.

Lemma 4.1. *Let $n \geq 1$ closed curves $\Gamma_1, \ldots, \Gamma_n$ of smoothness C^2 (possibly, with self-intersections or self-superpositions) be given on the plane. Then for any direction on the plane we can indicate an arbitrarily close direction $l = (\cos\varphi, \sin\varphi)$, such that if some straight line L parallel to the direction l is tangent to $\Gamma = \bigcup_{i=1}^n \Gamma_i$, then the point of tangency M of the straight line L with Γ is unique and any arc from Γ_i, tangent to L at the point M, has in M a curvature $k_i(M)$ different from zero[24].*

    Proof. We assume that some system of Cartesian coordinates is introduced on the plane, and the sense of rotation is chosen on all curves Γ_i. Consider the mapping f of the curve Γ_i on the circumference S^1 by parallel translation of the unit tangent vector to Γ_i to the origin of the coordinates. This mapping $f : \Gamma_i \to S^1$ is smooth and is irregular only at those points where the curvature k_i of the curve Γ_i is zero. This means that, by the Sard theorem, the points at Γ_i, where $k_i = 0$, have tangents, whose directions fill on S^1 a set of linear measure zero. Designate as E the set of those unit vectors on

---

[23] In [66], Subsection 3.2.1 (F), a similar result (without the proof) is also reported for C^∞-smooth isometric immersions.

[24] To understand the formulation of the lemma correctly, we should bear in mind that the point M is unique as a point of the plane, but as a point of the curve it can be a point of self-intersection of some curve Γ_i or a point of intersection of several curves from Γ. For instance, if the curve is a circumference twice covered, then any point on it will be considered two times as a point of the curve.

the plane, which are not collinear to the directions of the tangents to $\Gamma = \bigcup_{i=1}^{n} \Gamma_i$ at points of curvature $k = 0$. This set, on the strength of the above said, has full measure on S^1. Let us show that the set of those directions $\mathbf{e} \in E$, for which there is a straight line $L \parallel \mathbf{e}$, tangent to Γ in more than one point, has a linear measure zero on the circumference.

Let some vector $\mathbf{e} \in E$ be given. Introduce on the plane such a coordinate system (x, y) that $\mathbf{e}$ has the coordinates $\{\cos\varphi, \sin\varphi\}$, $\sin\varphi > 0$. Then an arbitrary straight line L, parallel to $\mathbf{e}$, has an equation

$$x = p + t\cos\varphi, \quad y = y_0 + t\sin\varphi, \tag{4.2}$$

where p and t are variables, and y_0 is some fixed value. Let L be tangent to Γ at two different points M_1 and M_2 belonging to the curves Γ_i and Γ_j (possibly, $i = j$). We parametrize these curves, respectively, in the neighbourhood of the points M_1 and M_2 by the arcs s and σ, and let $x = x_i(s)$, $y = y_i(s)$ and $x = x_j(\sigma)$, $y = y_j(\sigma)$ be the equations of the curves Γ_i and Γ_j, respectively, in the neighbourhood of M_i and M_j (we mean any one arc of the curve, if the curve has an intersection at this point). Then the condition of tangency of L with Γ_i and Γ_j leads to the equations

$$\begin{aligned}
p + t_1\cos\varphi = x_i(s), \quad y_0 + t_1\sin\varphi = y_i(s), \\
x_i'(s) = \varepsilon_1\cos\varphi, \quad y_i'(s) = \varepsilon_1\sin\varphi, \\
p + t_2\cos\varphi = x_j(\sigma), \quad y_0 + t_2\sin\varphi = y_j(\sigma), \\
x_j'(\sigma) = \varepsilon_2\cos\varphi, \quad y_j'(\sigma) = \varepsilon_2\sin\varphi. \\
\varepsilon_1 = \pm 1, \quad \varepsilon_2 = \pm 1.
\end{aligned} \tag{4.3}$$

At each curve Γ_i the points, at which the curvature $k \neq 0$, make either all the curve or no more than a countable set of open arcs $\gamma_i^{(m)} \subset \Gamma_i$, $1 \leq m \leq n_i$, where n_i is some natural number or ∞. We will show that at each pair of such arcs from Γ_i and Γ_j the set of solutions of the system (4.3) consists of isolated points. Indeed, independently on p, φ and y_0 from (4.3) we have

$$\varepsilon_1 x_i'(s) = \varepsilon_2 x_j'(\sigma), \tag{4.4}$$

$$\varepsilon_1 y_i'(s) = \varepsilon_2 y_j'(\sigma), \tag{4.5}$$

$$(x_i(s) - x_j(\sigma))y_j'(\sigma) = (y_i(s) - y_j(\sigma))x_j'(\sigma). \tag{4.6}$$

Consider the equations (4.4)–(4.6) as a system relative to s and σ. Let (s_0, σ_0) be some solution of this system. As the direction $\mathbf{e} \in E$, then $x_j''^2(\sigma_0) + y_j''^2(\sigma_0) \neq 0$ (from the condition that the curvature $k_j(\sigma_0) \neq 0$); say, $x_j''(\sigma_0) \neq 0$. Then from (4.4) we find the unique implicit smooth function $\sigma = \sigma(s), s \in (s_0 - \varepsilon, s_0 + \varepsilon)$, $\sigma(s_0) = \sigma_0$, with $\varepsilon_2 \sigma'(s_0) = \varepsilon_1 x_i''(s_0)/x_j''(\sigma_0)$. If the equations (4.4)–(4.6) are performed for some infinite set of points $(s_n, \sigma_n), n = 1, 2, \ldots$, with the limit point $(s_0, \sigma_0 = \sigma(s_0))$, then due to the uniqueness of the implicit function $\sigma = \sigma(s)$ all these points at sufficiently large n should lie on the line $\sigma = \sigma(s)$, and then, differentiating (4.5) and (4.6) at the points s_0 and $\sigma_0 = \sigma(s_0)$ with account for the equality $x_i'(s_0)y_j'(\sigma_0) = x_j'(\sigma_0)y_i'(s_0)$, we obtain

$$x_j''(\sigma_0)y_i''(s_0) = y_j''(\sigma_0)x_i''(s_0),$$

$$(x_i(s_0) - x_j(\sigma_0))y_j''(\sigma_0) - (y_i(s_0) - y_j(\sigma_0))x_j''(\sigma_0) = 0.$$

Together with (4.6) the latter equation yields the system relative to $x_i(s_0) - x_j(\sigma_0)$ and $y_i(s_0) - y_j(\sigma_0)$, which has the unique – zero – solution due to the distinction of the curvature Γ_j at the point $\sigma_0 = \sigma(s_0)$ from zero, i.e., $x_i(s_0) = x_j(\sigma_0)$ and $y_i(s_0) = y_j(\sigma_0)$, $M_1 = M_2$.

Therefore, irrespective of p, φ and y_0, the point (s_0, σ_0) satisfying (4.3)–(4.6) is isolated in each open set of the form $\gamma_i^{(m)} \times \gamma_j^{(n)}$, so that such points in this set and, consequently, in the entire $\Gamma_i \times \Gamma_j$, are no more than countable. Each straight line passing through these points with the simultaneous contingency of Γ_i and Γ_j, is determined by the equalities $\varepsilon_1 x_i'(s) = \varepsilon_2 x_j'(\sigma) = \cos \varphi, \varepsilon_1 y_i'(s) = \varepsilon_2 y_j'(\sigma) \sin \varphi$ and

$$x_i(s) - \frac{y_i(s) - y_0}{y_i'(s)} x_i'(s) = x_j(\sigma) - \frac{y_j(\sigma) - y_0}{y_j'(\sigma)} x_j'(\sigma) = p, \quad \text{consequently,}$$

such straight lines are also no more than a countable number. The lemma is proved.

A similar proposition also takes place for a set of straight lines emanating from one point, i.e., valid is the following:

Lemma 4.2. *Let $n \geq 1$ closed curves $\Gamma_1, \ldots, \Gamma_n$ of smoothness C^2 (possibly, with self-intersections or self-superpositions) be given on the plane. Then for any point of the plane we can indicate an arbitrarily close point M_0 such that if a ray L emanating from M_0 is* *tangent to $\Gamma = \bigcup_{i=1}^{n} \Gamma_i$, then the point of tangency M of the straight line L with Γ is unique and any arc from Γ_i, tangent to L at the point M, has in M a curvature $k_i(M)$ different from zero.*

The **proof** follows the same pattern as that of the previous lemma.

Let us turn to the proof of Theorem 4.1 in the case of smoothness $\Phi(\zeta)$ of class $C^m(\bar{D}_n), m \geq 2$. Let $\Gamma_1, \ldots, \Gamma_n$ be images of the circumferences making up the boundary of the domain D_n in the immersion $\Phi : \bar{D}_n \to E^2$. Using Lemma 4.1, we find the direction l, such that any straight line in E^2 parallel to l is either nowhere tangent to $\Gamma = \bigcup_{i=1}^{n} \Gamma_i$ or is tangent to Γ only at one point. Choose in E^2 the coordinate axes (x, y) such that the axis Oy be parallel to this direction l. Then the straight lines $x = const$ can be tangent to Γ at only one point, and the general number of tangency points will, evidently, be finite.

Generally speaking, we are interested in the case when the mapping $z = \Phi(\zeta)$ is many-sheeted (as the schlichtness of $\Phi(\zeta)$ implies that the mapping $z = x + iy = \Phi(\zeta)$ is already an embedding of $\bar{D}_n$ into $E^2 \subset E^3$), but the below construction is also suitable for the case of a univalent mapping $z = \Phi(\zeta)$. Draw in D_n through each point $\zeta_0 \in D_n$ a geodesic $\gamma(\zeta_0)$ at an angle $\theta = \dfrac{\pi}{2} - \arg \Phi'(\zeta_0)$ to the axis $O\xi$. The function $\Phi'(\zeta)$ is single-valued in D_n and is not equal to zero, so the direction of this geodesic in point ζ_0 is defined correctly. The angle θ is chosen such that the geodesic $\gamma(\zeta_0)$ during the mapping $z = \Phi(\zeta)$ passes into a segment of the straight line $x = const = Re\Phi(\zeta_0)$. It is easy to verify that along such a geodesic the mapping $z = \Phi(\zeta)$ is univalent (the coordinate $x = const$, and the coordinate y is monotone). Therefore, each geodesic like this is an analytical simple arc with the following potentialities: it is either nowhere tangent to ∂D_n, and consequently all the geodesic lies inside D_n with the exception of two endpoints on ∂D_n, or else it has one tangency point with ∂D_n, and then all the geodesic represents an arc with two endpoints on ∂D_n, broken up by the point of tangency with ∂D_n into two "halves" (at the endpoints, all geodesics $\gamma(\zeta)$ are transversal to ∂D_n). Further, on the strength of the uniqueness of the geodesic with a given direction at a given point, no two geodesics from the considered family γ intersect. As a result, we obtain that the entire domain D_n splits into geodesic families γ; what is more, the image of each such geodesic in D_n is a segment of a straight line $x = const$. Among geodesics of the family γ, of special significance are those tangent to ∂D_n. We will call them *boundary* geodesics. Points of their tangency with ∂D_n correspond to points of the local extremum of the function $x = Re\Phi(\zeta)$ on ∂D_n. Let N be the general number of points of the local extremum $Re\Phi(\zeta)$ on ∂D_n (this number N is finite on the strength of the initial choice of the direction of l in E^2; for the same reason, all extrema are strict). Let the extremum on ∂D_n in $N_1 \geq 2$ points of them be also a local extremum for the domain; and in $N_2 = N - N_1$ points the extremum is achieved only along ∂D_n. We will call these points *extreme* points of, respectively, *first* and *second* class. No geodesic of the family γ passes through extreme points of first class in ∂D_n (it degenerates into a point); a boundary geodesic passes through each extreme point of second class. Thus, the number of boundary geodesics is equal to N_2.

All boundary geodesics and extreme points of first class split D_n into a finite number of domains Ω_j. These domains are of two types: 1) the domain Ω_j is a curvilinear triangle bounded by an arc (the whole or "half" of it) of the boundary geodesic and by two arcs from ∂D_n emanating from an extreme point of first class; 2) the domain Ω_j is a curvilinear quadrangle bounded by two "opposite" arcs – parts of ∂D_n – and two arcs of two boundary geodesics.

On the plane (x, y), the first-type domain Ω_j in the mapping $z = \Phi(\zeta)$ is assigned to the domain bounded by lines of the form $x = const = c_1$, $y = f_1(x)$, $y = f_2(x)$, $f_1(x) < f_2(x), c_1 \leq x < c_2, f_1(c_2) = f_2(c_2)$ (if the value $x = c_2$ on ∂D_n is assigned to a point of the first-class local maximum) or else $x = const = c_2$, $y = f_1(x)$, $y = f_2(x)$, $f_1(x) < f_2(x), c_1 < x \leq c_2, f_1(c_1) = f_2(c_1)$ (if the value $x = c_1$ on ∂D_n is assigned to a point of the first-class local maximum). The second-type domain Ω_j on the plane (x, y) is assigned to a domain bounded by lines of the form $x = const = c_1$, $x = const = c_2$, $y = f_1(x)$, $y = f_2(x)$, $f_1(x) < f_2(x), c_1 \leq x \leq c_2$. Figure 3 shows differently hatched domains $\omega_j = \Phi(\Omega_j)$; first-class extrema are shown by asterisks; second-class points are designated by bolded dots.

Each domain $\omega_j = \Phi(\Omega_j)$ consists of segments $x = const$ with the monotone change of the *const* from some value c_1 to some value c_2; along each segment $x = const$ the mapping $z = \Phi(\zeta)$ is univalent, so in each closed domain Ω_j the mapping $z = \Phi(\zeta)$ is univalent in the large.

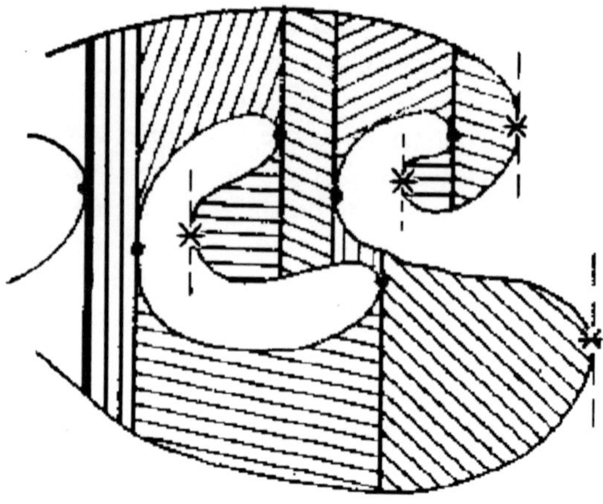

Figure 3. Domains $\Phi(\Omega_j)$ on the plane (x,y) (the case without self-superposition is shown).

Each second-class extreme point is the common boundary point of three domains Ω_j, and each first-class extreme point belongs to the boundary of only one domain Ω_j. Therefore, the number of first-type domains Ω_j is equal to N_1; of second-type domains, to $(3N_2 - N_1)/2$; and the number of all domains will be $(N_1 + 3N_2)/2$.

Let us now show that each domain Ω_j can be isometrically embedded into E^3 as a cylindrical surface, such that in the large we will obtain an embedding for D_n. In the case of a simply connected domain, the matter is simple: it is sufficient to construct the embedding of the domains Ω_j successively, with the regular junction by one generatrix. In the case of a multiply connected domain, embeddings Ω_j have to be matched in a more complex way: beginning an embedding with any domain of the form Ω_j, contiguous to the inner (for D_n) circumference Γ_i, after walking along Γ_i, we will have to match the embedding of the "last" domain of the form Ω_j both with its "preceding" neighbour and with the already embedded "first" domain. Let us adduce a construction, which will show that this matching process is quite feasible.

Let us mark all N points of first- and second-class extrema in D_n. At the boundary of each domain Ω_j there are two such points: if a domain is of first type, then one of the points is of first class and the other of second class, while in a second-type domain both these points are of second class. Let us join these marked points by a simple Jordan arc in Ω_j. As a result, we obtain some graph G, the vertices of which are the marked N points of the local extremum of the function $x = Re\Phi(\zeta)$ on ∂D_n, and the edges are Jordan arcs constructed in each domain Ω_j. As the domains Ω_j are not overlapped by one another, the edges intersect only in the vertices. Only one edge emanates from the vertices corresponding to first-class extreme points, while three edges emanate from the second-class vertices. In the case of a simply connected domain D_1 the graph G contains no cycles, and in the case of the multiply connected domain D_n, $n \geq 2$, the graph G will contain exactly $(n-1)$ independent cycles.

Let us introduce the coordinates X, Y, Z in the space E^3 and take the above constructed graph (with up to the isomorphism) for the projection of the sought-for cylindrical embedding D_n into E^3 on the plane XZ. The concrete construction of this projection will be carried out as follows: on the plane XZ we take a graph G_ε without self-intersections, with rectilinear edges of no more than ε long, isomorphic to the graph G (the value of ε will be made concrete a little later). Let δ be the least distance between the vertices of the constructed graph G; assume $t = min(\delta, \frac{\varepsilon}{2})$. We describe around each vertex A_i two concentric circles of radii $r = t/4$ and $r = t/8$, respectively.

Now note that of three domains of the form Ω_j, contiguous to one second-class vertex M, one domain contains as part of its boundary the whole boundary geodesic passing through the point M on ∂D_n (namely, the domain Ω_j, which contains no arc of circle from ∂D_n passing through M), and two others contain on their boundary by one "half" each of the boundary geodesic, into which it is broken down by the point of tangency with ∂D_n. We call the first domain a "large" domain and two others "small" domains.

Let us now change the graph G_ε as follows (Figure 4). Consider the circles $K^i_{t/8}$ and $K^i_{t/4}$, circumscribed around an i-th vertex of second class $A_i \in G_\varepsilon$. In these circles, the edges emanating from A_i form three radii $A_iC^i_1$, $A_iC^i_2$ and $A_iC^i_3$. Let $A_iC^i_1$ be the radius in $K^i_{t/4}$ and $A_iB^i_1 \subset A_iC^i_1$, the radius in $K^i_{t/8}$, corresponding to the "large" domain of the form Ω_j, contiguous to A_i; $A_iC^i_2$, $A_iB^i_2 \subset A_iC^i_2$ and $A_iC^i_3$, $A_iB^i_3 \subset A_iC^i_3$ have a corresponding meaning for the "small"

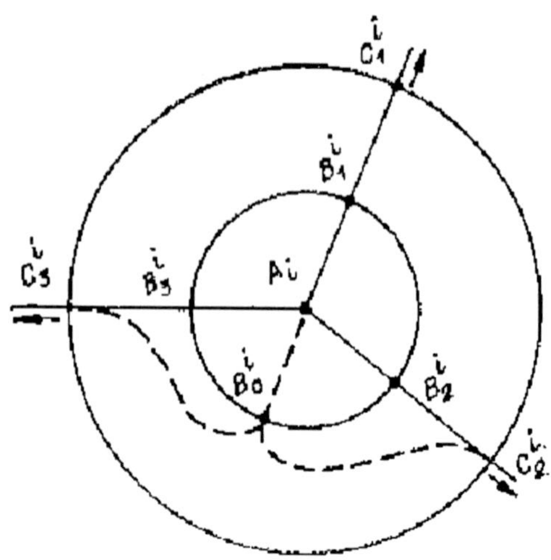

Figure 4. Deformation of the graph G_ε in the neighbourhood of the vertex A_i: the radii $A_i C_2^i$ and $A_i C_3^i$ are replaced by the dashed lines $A_i B_0^i C_2^i$ and $A_i B_0^i C_3^i$.

domains of the form Ω_j, contiguous to A_i. Let us extend the radius $B_1^i A_i$ to the diameter $B_1^i A_i B_0^i$ of the circle $K_{t/8}$, and in the two other edges emanating from A_i, let us replace their parts – the radii $A_i C_2^i$ and $A_i C_3^i$ – by the segment $A_i B_0^i$ – the extension of $B_1^i A_i$ up to the diameter, and C^∞-smoothly contiguous to $A_i B_0^i$ (at the point B_0^i), to $A_i C_2^i$ (at the point C_2^i) and to $A_i C_3^i$ (at the point B_0^i) by the Jordan arcs $B_0^i C_2^i$ and $B_0^i C_3^i$ located in the circle $K_{t/4}^i \setminus K_{t/8}^i$ and not intersecting. As the result, we have a new graph G_ε', in which the edges are C^∞-smooth Jordan arcs without mutual intersection; moreover, the edge corresponding to the "large" domain Ω_j, adjacent to the considered second-class point, adjoins C^∞-smoothly (and even analytically) two other edges emanating from the same point and respective "small" domains of the form Ω_j. Outside of the circles $K_{t/4}^i$ all the edges of the graph G_ε' are as before segments of straight lines. We indicate the least distance between these segments as λ.

Each domain $\Omega_j \subset D_n$ is now embedded into E^3 as follows. Take in the graph G_ε' the edge $A_{j_1} A_{j_2}$ corresponding to the domain Ω_j. Let its equation be $X = X_j(s)$, $Z = Z_j(s)$ (relative to the Euclidean length s of the edge $A_{j_1} A_{j_2}$). The domain Ω_j on the plane (x, y) is assigned a domain lying between the straight lines $x = c_1^j$ and $x = c_2^j$, $c_1^j < c_2^j$, and the curves of the form $y = f_1^j(x)$, $y = f_2^j(x)$, $c_1^j < x < c_2^j, f_1^j(x) < f_2^j(x)$. If the length of the edge $X = X_j(s)$, $Z = Z_j(s)$ in the graph G'_ε is equal to $c_2^j - c_1^j$, we do not change this edge any more; but if the length of this edge is smaller than $c_2^j - c_1^j$, we change this edge at its straight segment l outside the circles of the form $K_{t/4}^i$ such that the edge remain C^∞-smooth, be not removed from l to a distance of more than $\lambda/4$ and that the length of the entire edge be equal to $c_2^j - c_1^j$ (we choose ε such that the length of the edge be not greater than $c_2^j - c_1^j$ at any j). Let the new graph and new edges in it have the previous designations – G'_ε and $X = X_j(s)$, $Z = Z_j(s), 0 \leq s \leq c_2^j - c_1^j$. Then the embedding of the domain Ω_j into E^3 will be defined by the following relations:

$$X = X_j(s), \quad s = x(\zeta) - c_1^j, \quad Y = y(\zeta), \quad Z = Z_j(s),$$

in the assumption that the point $X_j(0)$, $Z_j(0)$ corresponds to a point on ∂D_n with $x = \operatorname{Re}\Phi(\zeta) = c_1^j$. By the construction, all the graph G'_ε – the projection to XZ of the indicated immersion of D_n into E^3 – has no self-intersections outside the circles $K_{t/8}^i$, and "small" domains of the form Ω_j contiguous to one vertex of second class A_i have projections to the plane XZ, coinciding in $K_{t/8}^i$. But in space at a sufficiently small t they are assigned various points, as their coordinates Y will be different (on the plane (x, y)). In the neighbourhood of a second-class extreme point, two "small" domains Ω_j are defined by the following relations: for one domain $x \geq c_1^j, f_1^j(x) < y < f_2^j(x)$, for the other $x \geq c_1^j, f_3^j(x) < y < f_4^j(x) < f_1^j(x), f_1^j(c_1) = f_4^j(c_1)$ or else $x \leq c_1^j, f_1^j(x) < y < f_2^j(x)$ and $x \leq c_1^j, f_3^j(x) < y < f_4^j(x) < f_1^j(x), f_1^j(c_1) = f_4^j(c_1)$. Therefore, the constructed mapping $D_n \to E^3$ is an embedding in the large if t is chosen to be sufficiently small, such that the above indicated inequalities $f_4^j(x) < f_1^j(x)$ be satisfied along the radius $A_i B_0^i$ at all $i \leq N_2$, which is, evidently, met. Besides, it is easy to verify that in interior points the constructed embedding will be of smoothness class C^∞, and the boundary curves are of the same smoothness as the mapping $\Phi(\zeta)$ in $\bar{D}_n$ (or the smoothness C^∞, if $\Phi(\zeta)$ is analytical). The theorem is proved.

Remark 4.1. By the condition of the theorem, the mapping $\Phi(\zeta)$ for the isometric immersion of the domain D_n into E^2 was of smoothness no less than $C^2(\bar{D}_n)$. But if the metric (4.1) has in $\bar{D}_n$ the smoothness of only class $C^{0,\alpha}, 0 < \alpha \leq 1$, the smoothness of class C^2 cannot be guaranteed for $\Phi(\zeta)$. Then we do as follows. In this case, we have $\Phi(\zeta) \in C^{1,\alpha}$. Let us take an arbitrary point $\zeta_0 \in \partial D_n$; let (x_0, y_0) be its image in E^2 in the immersion $z = \Phi(\zeta)$. Let $(x(s), y(s))$ be the equation of the arc of the boundary $\partial\Phi(D_n)$ in the neighbourhood of the point (x_0, y_0). This boundary is locally Jordan and smooth; therefore, the arc through the point (x_0, y_0) divides the sufficiently small circle with its centre in (x_0, y_0) into two parts, C^1-diffeomorphic to a semi-circle, such that points from the image $\Phi(D_n)$ lie in one part, and there are no such points from the neighbourhood of ζ_0 in the other (it is useful to bear in mind that the considered arc projects uniquely to its tangent at its sufficiently small dimensions). Therefore, we can expand the image $\Phi(D_n)$ in the plane (x, y) by adding to each point from $\partial\Phi(D_n)$ the semi-circle in which there were earlier no points from the image $\Phi(D_n)$. As $\Phi'(\zeta) \neq 0$ in $\bar{D}_n$ and ∂D_n is compact, then such an expansion can be made by adding a finite number of semi-circles, such that all the previous boundary would prove to be an inner curve. After that, it is now easy to draw a curve of class C^∞, for which there is the direction l with the required property of tangency with the boundary in a finite number of points lying by no more than one on the tangent to the boundary of the straight line. Therefore, we obtain an embedding of this domain D_n as part of the embedding of a broader domain. Thereby, the theorem is also valid for metrics of class $C^{0,\alpha}, 0 < \alpha \leq 1.\bullet$

Remark 4.2. In concrete cases there is no need to construct the graph G'_ε by the general scheme described in the proof of Theorem 4.1. The very position of the boundary geodesics in D_n sometimes suggests the form of the graph G'_ε; it is only important that an edge corresponding to the "large" domain be matched to a desired smoothness with edges corresponding to the "small" domains adjacent with this "large" domain.$\bullet$

To illustrate this remark and, generally, the theorem, let us consider the following:

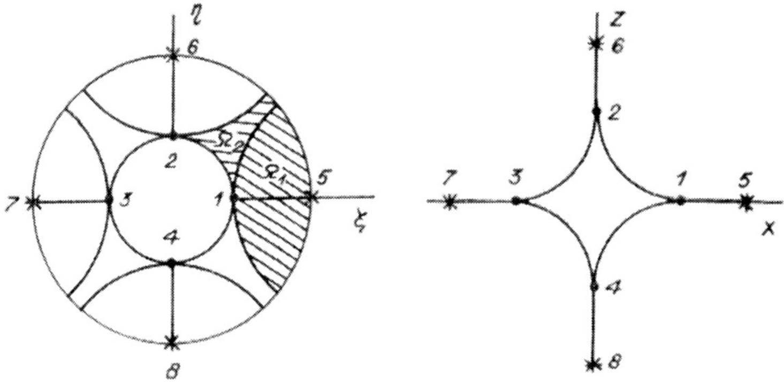

Figure 5. On the left-hand side – the domains Ω_j and the graph G in the ring; on the right-hand side – the graph G_ϵ on the plane XZ.

Example. In the ring $\bar{D}_2 : r \leq |\zeta| \leq R$, consider the metric

$$ds^2 = 4(\xi^2 + \eta^2)(d\xi^2 + d\eta^2).$$

Its immersion into E^2 is given by the mapping $z = \Phi(\zeta) = \zeta^2$, and the image in immersion represents a twice covered ring $r^2 \leq |z| \leq R^2$. The segments $x = \pm r^2$ tangent to the circumference $|z| = r^2$ are assigned four boundary geodesics $\xi^2 - \eta^2 = \pm r^2$ and four extreme points of second class: A_1 with the coordinates $(\xi_1 = r, \eta_1 = 0)$, A_2 with $(\xi_1 = 0, \eta_2 = r)$, A_3 with $(\xi_3 = -r, \eta_3 = 0)$ and A_4 with $(\xi_1 = 0, \eta_4 = -r)$. The graph G can be taken in the form of arcs of circle $\xi^2 + \eta^2 = r^2$, joining consecutively four points of second class on it, and four radii joining four points of second class with respective points of first class $A_5(\xi_5 = R, \eta_5 = 0)$, $A_6(\xi_6 = 0, \eta_6 = R)$, $A_7(\xi_7 = -R, \eta_7 = 0)$ and $A_8(\xi_8 = 0, \eta_5 = -R)$. The number of domains Ω_j is 8. We take the graph G_ϵ in the plane XZ in the form of an "astroid with whiskers" (Figure 5):

$$X^{2/3} + Z^{2/3} = \left(\tfrac{4}{3}r^2\right)^{2/3},$$

$$\tfrac{4}{3}r^2 \leq X \leq R^2 + \tfrac{r^2}{3}, \ Z = 0; \ X = 0, \ \tfrac{4}{3}r^2 \leq Z \leq R^2 + \tfrac{r^2}{3},$$

$$-R^2 - \tfrac{r^2}{3} \leq X \leq \tfrac{4}{3}r^2, \ Z = 0; \ X = 0, \ -R^2 - \tfrac{r^2}{3} \leq Z \leq -\tfrac{4}{3}r^2.$$

I.Kh. SABITOV

The length of each of the arcs of the astroid is equal to $2r^2$, i.e., to the difference between $x = c_1 = -r^2$ and $x = c_2 = r^2$; the length of each "whisker" is also equal to the difference between the neighbouring extreme values $x = \pm r^2$ and $x = \pm R^2$. The embedding into E^3 of the ring $\bar{D}_2 : r \leq |\zeta| \leq R$ with the metric (4.6) is given by the following relations:

$$\text{domain } \Omega_1 : \quad \xi^2 - \eta^2 \geq r^2, \xi > 0,$$

$$X = \xi^2 - \eta^2 + \frac{r^2}{3} \left(= X(s) = s + \frac{4}{3}r^2 = x(\xi, \eta) + \frac{1}{3}r^2 \right),$$

$$Y = 2\xi\eta, \quad Z = 0, \quad 0 \leq s \leq R^2 - r^2;$$

$$\text{domain } \Omega_2 : \quad -r^2 \leq \xi^2 - \eta^2 \leq r^2, \xi > 0, \ \eta > 0,$$

$$X = \frac{\sqrt{2}}{3r}(r^2 + \xi^2 - \eta^2)^{3/2} \left(= X(s) = \frac{4}{3}r^2 \left(1 - \frac{2s}{4r^2}\right)^{3/2} = \right.$$

$$\left. \frac{4r^2}{3} \left(1 - \frac{r^2 - x(\zeta)}{2r^2}\right)^{3/2} \right), \quad Y = 2\xi\eta,$$

$$Z = \frac{\sqrt{2}}{3r}(r^2 - \xi^2 + \eta^2)^{3/2} \left(= Z(s) = \frac{4}{3}r^2 (\frac{2s}{4r^2})^{3/2} = \right.$$

$$\left. \frac{4r^2}{3} \left(\frac{r^2 - x(\zeta)}{2r^2}\right)^{3/2} \right), \quad 0 \leq s \leq 2r^2;$$

etc.,

$$\text{domain } \Omega_8 : \quad -r^2 \leq \xi^2 - \eta^2 \leq r^2, \xi > 0, \ \eta < 0,$$

$$X = \frac{\sqrt{2}}{3r}(r^2 + \xi^2 - \eta^2)^{3/2}, \quad Y = 2\xi\eta,$$

$$Z = -\frac{\sqrt{2}}{3r}(r^2 - \xi^2 + \eta^2)^{3/2}.$$

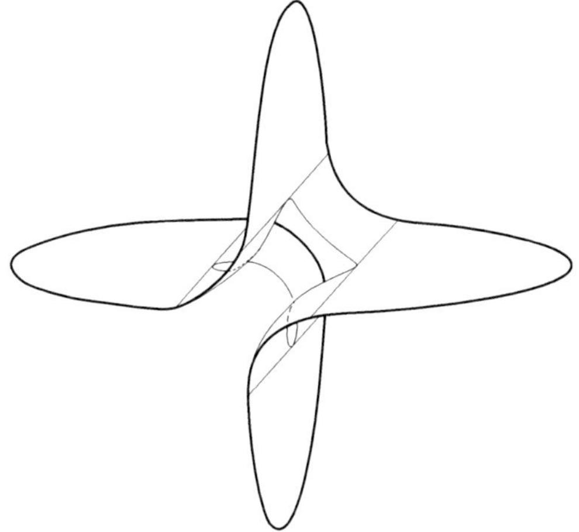

Figure 6

As the result we obtain a cylindrical surface without self-intersections, everywhere analytical, besides images of boundary geodesics; the general smoothness is of class $C^{1,1/2}$, but it appears easy to bring the smoothness up to C^∞ (Figure 6).

Remark 4.3. From the proof of Theorem 4.1 it is seen that D_n can be embedded into E^3 in the form of a cylindrical surface placeable in a circular cylinder of any prescribed arbitrarily small radius.●

Remark 4.4. Using Lemma 4.2, by analogous reasoning it is shown that D_n can be isometrically embedded into E^3 as a domain on a conical surface. Such embeddings are sometimes possible for domains, for which there are no embeddings and even immersions in the form of cylindrical surfaces. For instance, on a plane with a punctured point $(0,0)$ let us give a metric

$$ds^2 = \rho^{2n-2}(d\xi^2 + \eta^2), \rho^2 = \xi^2 + \eta^2, \zeta - \zeta \mid i\eta.$$

When reducing it to the standard form by the mapping $z = \dfrac{\zeta^n}{n}$ we obtain that it corresponds to the metric of an n-multiply covered punctured Euclidean plane. This metric cannot be embedded as a cylindrical surface, but it admits an embedding as a conical surface with the vertex in the centre of a unit sphere and the directrix in the form of a spherical curve without self-intersections of length $2\pi n$.$\bullet$

Open question 4.1. Surfaces obtained by the described method will be everywhere analytical in the case of analyticity of the boundary curves of a many-sheeted domain, besides the points of generatrices corresponding to the boundary geodesics. If the boundary of the domain Ω is not analytical, then we can slightly extend this domain and obtain a new domain with an analytical boundary, so that the initial open domain will prove to be embedded into R^3 as a surface analytical everywhere besides the boundary generatrices for the new extended domain. The question is whether for an l. E. metric given in the simply connected domain D we can construct its embedding into R^3, which would be an analytical surface in the open domain.$\bullet$

4.2.2 *Proof of Theorem 4.2*

With account for the results of Subsection 2.4.4, it can be shown that the mapping $z = \Phi(\zeta)$ of the ring $\bar{D}_2 : r \leq |\zeta| \leq R$ with an l. E. metric into E^2 in the case of many-valuedness $\Phi(\zeta)$ has the following form:

$$z = \Phi(\zeta) = A \ln \zeta + \Phi_0(\zeta), \, A \neq 0 \qquad (4.7)$$

or

$$z = \Phi(\zeta) = \zeta^{\alpha/2\pi} \Phi_0(\zeta), \, \frac{\alpha}{2\pi} \notin \mathbf{Z}, \qquad (4.8)$$

where Φ_0 is a single-valued holomorphic function.

In the case of (4.7) we have a cylindrical type of many-valuedness. By rotating the axes x, y we can achieve that in the new axes (we leave for them the previous notation x, y) the mapping $z = \Phi(\zeta)$ has the form

$$z = x + iy = iB \ln \zeta + \Phi_0(\zeta)$$

with the real coefficient $B \neq 0$. Then the mapping

$$X = B \cos \frac{x(\xi, \eta)}{B}, \; Y = y(\xi, \eta), \; Z = B \sin \frac{x(\xi, \eta)}{B}$$

is, evidently, single-valued and yields the isometric immersion D_2 into E^3 in the form of a domain (probably, many-sheeted) on a circular cylinder $X^2 + Z^2 = B^2$.

In the case of (4.8) we have a conical type of many-valuedness. For the description of immersion, we need to distinguish between two possibilities here.

a) The function $\Phi_0(\zeta) \neq 0$ in $\bar{D}_2$. In this case, $\Delta arg\Phi(\zeta)$ – the increment of the argument $\Phi(\zeta)$ – is the same in walking along any closed contour homotopic to the boundary circumferences of the ring. We assume

$$\varkappa = \frac{1}{2\pi}\Delta\Phi(\zeta) \notin \mathbf{Z}$$

(the walking along any boundary circumference). Let $[a]$ denote the integral part of the number a. Consider the mapping $\bar{D}_2 \to E^3$ defined by the relations

$$X = \mu|\Phi(\zeta)|\cos\frac{arg\Phi(\zeta)}{\mu},$$
$$Y = \sqrt{1 - \mu^2}|\Phi(\zeta)|,$$
$$Z = \mu|\Phi(\zeta)|\sin\frac{arg\Phi(\zeta)}{\mu}, \mu = \frac{\varkappa}{1 + [|\varkappa|]},$$

and being, evidently, single-valued and of the same regularity in $\bar{D}_2$ as $\Phi(\zeta)$. This mapping does yield the immersion of D_2 into E^3 in the form of some domain (possibly, many-sheeted) on a circular cone $X^2 + Z^2 = (\mu^2/(1 - \mu^2))Y^2$.

b) The function $\Phi_0(\zeta)$ becomes (simple) zero in one or several points of the domain $\bar{D}_2$ (becoming zero in an infinite number of points is impossible in view of the condition $\Phi'(\zeta) \neq 0$ in $\bar{D}_2$). In this case the previous mapping of $D_2 \to E^3$ is many-valued in the neighbourhoods of the zeros of the function $\Phi_0(\zeta)$; an analogous mapping transferring any other point $z_0 = x_0 + iy_0$ to the vertex of the cone is not suitable, either, as $\Delta arg(\Phi - z_0)$ has various values for the curves, which are, however, homotopic between themselves. Let us describe how we can construct an isometric immersion $D_2 \to E^3$ in this case (this technique is also suitable for the previous case).

We can assume that $\Phi_0(\zeta)$ goes to zero only in interior points of the ring D_2, because the domain of setting an l. E. metric can always be made slightly wider, the way it is done, for instance, at the end of the proof of Theorem 4.1. We draw in D_2 all geodesics, along which $arg\, \Phi(\zeta) = const$. Such geodesics on the plane (x, y) are assigned segments lying on the rays $arg\, z = const$. The position of these geodesics in D_2 in total does not depend on the choice of the branch of the many-valued function $\Phi(\zeta)$, because at a change of the branch during the complete circulation around $\zeta = 0$ or around zero $\Phi_0(\zeta)$ the value of the argument along the considered geodesic would change by only the same constant value. Along any geodesic the mapping $z = \Phi(\zeta)$ is univalent, so the geodesic emanating from one zero of the function $\Phi(\zeta)$ does not come to the other zero of $\Phi(\zeta)$; for the same reason, any two geodesics (of the considered family) may intersect only at the zero of the function $\Phi(\zeta)$.

Consider all geodesics emanating from some zero of the function $\Phi(\zeta)$. Their union cannot cover the entire ring D_2 (indeed, otherwise, on the strength of the above mentioned properties of the considered geodesics the ring D_2 would have been homeomorphically mapped on the plane (x, y) into some simply-connected domain star-shaped with respect to the point $z = 0$). Note further that the geodesic $arg\, \Phi = const$ emanating from a zero of $\Phi(\zeta)$ (by condition lying inside D_2) can have common points simultaneously with both boundary circles $|\zeta| = r$ and $|\zeta| = R$ only provided it is tangent to one of them (as always, we mean not the complete geodesic passing *through* the point of zero of the function $\Phi(\zeta)$ and touching this direction, but part of the geodesic, along which $arg\, \Phi = const$).

We now take some concrete geodesic $\gamma_{c_0} : arg\, \Phi(\zeta) = c_0$ emanating from the considered zero ζ_0 of the function $\Phi(\zeta)$. Assume that in its extension into $\bar{D}_2$ it has no common points with the circumference, say, $|\zeta| = r$. Then its final point M_{c_0} belongs to the circumference $|\zeta| = R$ (as the mapping $z = \Phi(\zeta)$ is bounded and is univalent along the geodesic, the geodesic cannot be extended infinitely and at the same time cannot end in an interior point or return to ζ_0). At a continuous and monotone change of $const = c$ near c_0 the last point M_c of the geodesic γ_c will monotonously (but not necessarily continuously) move along the circumference $|\zeta| = R$. As during the change of $const$ from c_0 up to $c_0 + 2\pi$ the geodesic $arg\, \Phi(\zeta) = const$ emanating from ζ_0 should return to the initial position γ_{c_0}, then during the change of $const$ from c_0 we would have the first value $const = c_1 > c_0$, at which the geodesic $\gamma_{c_1} : arg\,\Phi(\zeta) = c_1$ would either for the first time touch the circumference $|\zeta| = r$ and then would go up to the circumference $|\zeta| = R$ or else would touch the circumference $|\zeta| = R$ for the last time and would go further up to the circumference $|\zeta| = r$. In any case, we would have the arc γ of the geodesic with $arg\,\Phi(\zeta) = const$, going from the circumference $|\zeta| = R$ up to the circumference $|\zeta| = r$ and passing through no zero of the function $\Phi(\zeta)$.

Cut D_2 along this simple arc γ. Its images in the plane (x, y) at the mapping $z = \Phi(\zeta)$ will be two segments of equal length $l_1 :$ $arg\,z = const$ and $l_2 = const + \alpha + 2n\pi, n \in \mathbf{Z}$, not passing through the origin of coordinates and beginning at the same distance from the origin of coordinates. The rays on which these segments are situated meet in the origin of coordinates at an angle α_0, where $\alpha_0 = 2m\pi + \alpha$, $|\alpha_0| < 2\pi$, $m \in \mathbf{Z}$. We draw through the origin of coordinates the straight line b, which is the bisectrix of the angle between the given rays. This straight line will split the plane z into two half-planes. We turn one half-plane around b till it coincides with the other half-plane. Then the segment l_1 will align with the segment l_2, and we will have the nonsmooth realization of D_2 in the form of a domain on half-planes superposed on each other. It is clear that now we will need to smooth this mapping on the bend of the plane (x, y) along the segments of the straight line b, which got into each sheet of a possibly many-sheeted domain $\Phi(D_2)$, and along the coincident segments l_1 and l_2, which is done by substitution of the cylindrical surface having a directrix of sufficiently large curvature on the generatrices b and $l_1 = l_2$ for the narrow strip of the surface along b and along $l_1 = l_2$. The analytic description of this deformation of strips of the plane, superposed one on the other, is given by the following construction.

Let two halves of the domain $T : -a \le x \le a, f_1(x) \le y \le f_2(x)$ situated on the plane xy be isometrically mapped in E^3 with the coordinates (X, Y, Z) into two domains T_1 and T_2 on the half-plane $YZ, Z \ge 0$, where $T_1 : 0 \le Z \le a, X = 0, f_1(Z) \le y \le f_2(Z)$, and $T_2 : 0 \le Z \le a, X = 0, f_1(-Z) \le y \le f_2(-Z)$, i.e., the part of the domain T, where $x \ge 0$, passed by the motion into the domain T_1, and the part, where $x \le 0$, passed into T_2. We introduce the functions

$$X = g_1(Z) = \begin{cases} g_\varepsilon(Z) > 0 & \text{for } \varepsilon/2 < Z < \varepsilon < a \\ 0 & \text{for } Z = \varepsilon/2 \text{ and } \varepsilon \le Z \le a. \end{cases}$$

and

$$X = g_2(Z) = \begin{cases} -g_\varepsilon(Z) > 0 & \text{for } \varepsilon/2 < Z < \varepsilon < a \\ 0 & \text{for } Z = \varepsilon/2 \text{ and } \varepsilon \leq Z \leq a. \end{cases}$$

We choose the function $g_\varepsilon(Z)$ such that the curve $\Gamma : (g_1(Z), 0, Z)$ $\cup (g_2(Z), 0, Z)$ be C^∞-smooth and that

$$\int_{\varepsilon/2}^{\varepsilon} \sqrt{1 + g_\varepsilon'^2(t)}\, dt = \varepsilon.$$

The monotone function $s = s(Z) = \int_{\varepsilon/2}^{Z} \sqrt{1 + g_\varepsilon'^2(t)}\, dt$ has the following properties: $s(\varepsilon/2) = 0$, $s(\varepsilon) = \varepsilon, s(Z) = Z, Z \geq \varepsilon$. In accordance with this, the cylindrical surfaces $S_1 : x = g_1(Z)$, $\varepsilon/2 \leq Z \leq a, f_1(s(Z)) \leq Y \leq f_2(s(Z))$ and $S_2 : x = g_2(Z)$, $\varepsilon/2 \leq Z \leq a, f_1(-s(Z)) \leq Y \leq f_2(-s(Z))$ together give a C^∞-smooth immersion into E^3 of the domain T.

Thus, we obtained an isometric immersion of D_2 into E^3 as a surface consisting of several superposed plane pieces, joined by cylindrical strips, the directrix of which has the form of the curve $y = \pm x^n \sqrt{\varepsilon - x}$, $\varepsilon > 0, n > 1$. Theorem 4.2 is proved.

Remark 4.5. The fact that the "folding" of the development of a conical surface into a cone is not a trivial problem can be demonstrated using a model shown in Figure 7. A criterion of its structure is that the extensions of the edges of the development intersect in an interior point of the domain M, of which the surface consists. If we try to "glue" or identify the edges of the development in the usual way, by directly hurling them together, we can feel that paper does not "yield", and a "beak" emerges in the region of the point M, in the neighbourhood of which the surface tends to lose its smoothness.●

Remark 4.6. It follows from the expounded method that in cases 1) and 2a) the embeddability of D_2 into E^3 can, apparently, be established by using the same technique as in the proof of Theorem 4.1, but in case 2b) the consideration of the existence of the embedding should require new approaches.●

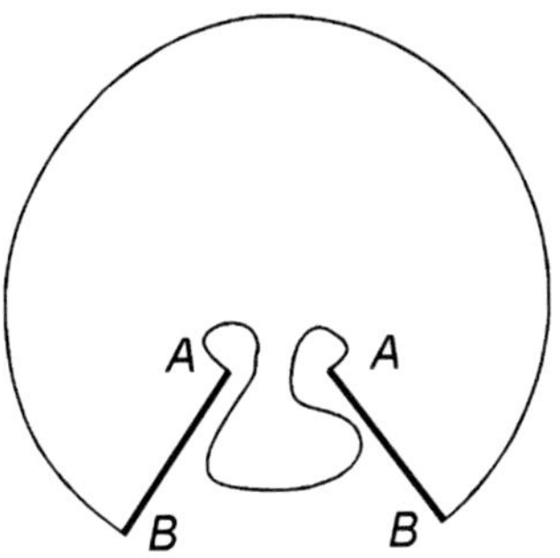

Figure 7

*4.2.3   An example of an unimmersible triply connected domain with
a flat metric*

In this Subsection we present an example of an l. E. metric given in a
triply connected domain and not immersible into E^3 in the form of a
C^1-smooth torse. As we have already established in Chapter 3, C^1-
smooth torses have the geometry usual for developable surfaces. Recall that this implies that a rectilinear segment (a generatrix) passes
through each of its points, along which segment the tangent plane to
the surface is stationary. Further, if a point has no plane neighbourhood, the generatrix passing through it is unique and goes from edge
to edge of the surface, and all along its length none of its points has
a plane neighbourhood. If, however, two generatrices l_1 and l_2 pass
through any point, then, first, this point has a plane neighbourhood
and, second, all generatrices intersecting with l_1 or l_2 lie on the same
plane, so that the corresponding part of the surface is a plane domain.

To construct the sought-for example, first we will prove the following lemma.

Lemma 4.3. *Let a rectangle $A_1A_2D_2D_1$ be given on the Euclidean
plane (see Figure 8) and let $|A_1A_2| = |B_1B_2| = 2R$, $|A_1B_1| =
|A_2B_2| = h$, $|B_1M_1| = |M_2B_2| = R/2$. We identify the segments
A_1B_1 and A_2B_2. Let a line l be drawn through some point $M_0 \in
[M_1, M_2]$. This line is the complete shortest arc in the new metric
(i.e., it comes with its both ends to the edge and yields the short-*

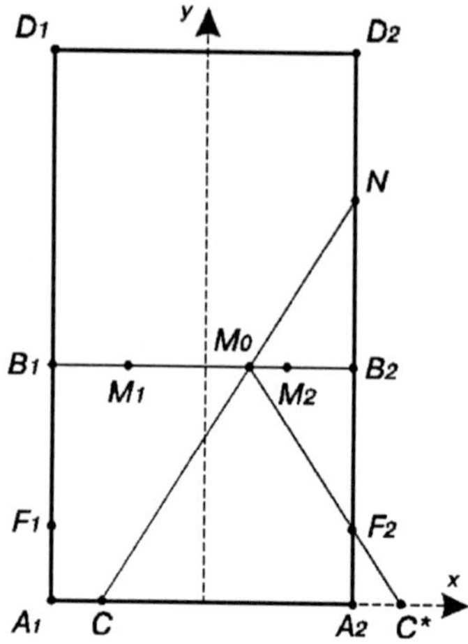

Figure 8

*est distance for any pair of its points) and connects the point M_0
with some point C of the segment $A_1 A_2$. Then, if the lengths of the
segments $B_1 D_1$ and $B_2 D_2$ are smaller than $h/2$, the shortest arc l
intersects the side $D_1 D_2$ of the rectangle.*

Proof. Let a system of coordinates be introduced in the rectangle
the way it is indicated in Figure 8. Without loss of generality, we can
consider that the point M_0 has the coordinate x_0 with the condition
$0 \leq x_0 \leq R/2$. Let the shortest arc l with the equation $y = h +
k(x - x_0)$ pass through the point M_0 and let it for definiteness come
to the edge at some point $N \in (B_2 D_2)$, so that $k > 0$ and the point of
intersection of l with $A_1 A_2$ have the coordinates $x_C = x_0 - h/k, y_C =
0$. The segment $M_0 C$ of this shortest arc does yield the shortest
distance between the points M_0 and C in that and only that case
when $|M_0 C| \leq |M_0 C^*|$, where C^* is a point with the coordinates
$(2R + x_C, 0)$ (otherwise, we will have another shortest arc $[M_0 F_2] \cup
[F_2 C^*] = [M_0 F_2] \cup [F_1 C]$, see Figure 8). This gives us the following
condition for k:

$$x_0 - x_C \leq x_{C^*} - x_0 \Leftrightarrow h/k \leq R \Leftrightarrow k \geq h/R.$$

Thus, the shortest arc passing through M_0 and coming to $B_2 D_2$ is obliged to have the slope $k \geq h/R$. Then the "lowest" point N, in which the shortest arcs from the points $M_0 \in [M_1 M_2]$ intersect $(B_2 D_2)$, is defined as the intersection of the very possible "sloping" shortest arc passing through the point M_2, which gives us the ordinate $y_N = \dfrac{3}{2}h$. Therefore, if the length of $B_2 D_2$ is not greater than $h/2$, then any shortest arc of the considered type, passing through any point from $[M_1 M_2]$, will intersect the side $D_1 D_2$. The lemma is proved.

Remark 4.7. In reality, for the admissible length L of the side $B_2 D_2$ with the proposition of the lemma preserved we can give a better estimate of the form of $L \leq ah$, $a > 1/2$. It is obtained in the following way. As the segment CN is the shortest arc between its ends, any path from N to C has the length not smaller than $L_1 = |CN|$. In particular, the path $NB_2 \cup B_2 C^* = NB_2 \cup B_1 C$ has the length $L_2 = k(R - x_0) + \sqrt{h^2 + (R + x_0 - h/k)^2}$, which should be not smaller than L_1. This condition gives us the inequality

$$hk(R - x_0) + 2R(\frac{h}{k} - x_0) \leq k(R - x_0)\sqrt{h^2 + (R + x_0 - h/k)^2},$$

whose solution leads to some lower estimate for k, whence it could be deduced that the lemma is also valid if the length of $B_2 D_2$ is greater than $h/2$ (but, certainly, with some upper bound).•

Consider a plane domain shown in Figure 9. If we identify the sides $A_1 B_1$ and $A_2 B_2$, then we obtain a domain with an l. E. metric, and the boundary of the domain will be the closed lines $A_1 A_2$ and $B_1 C_1 D_1 D_2 C_2 B_2$. Let the arcs $B_1 C_1$ and $B_2 C_2$ be semi-circles of radius r. Let the length of $A_1 A_2$ be equal to $2R$, the length of $A_1 B_1$ be h, and the length of $C_2 D_2$ be strictly less that $h/2$. Assume that this domain is realized in E^3 in the form of some C^1-smooth torse S. The segment $B_1 B_2$ of length $2R$ should be realized on S in the form of some closed smooth curve γ, on which the arc corresponding to the segment $M_1 M_2$ of length R can, evidently, not consist totally of planar points and, therefore, there is at least one point $\gamma(M_0)$ on it, which has no plane neighbourhood and through which the unique generatrix l passes, which is simultaneously the complete shortest arc.

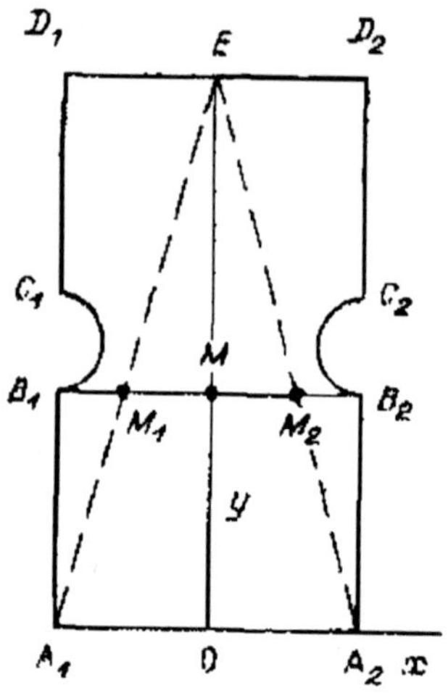

Figure 9

This shortest arc cannot run along γ and cannot be tangent to γ (as only one geodesic passes in this direction), therefore, it intersects γ, so one of its ends should be on the image $A_1 A_2$. If the radius r of the semi-circles $B_1 C_1$ and $B_2 C_2$ is chosen to be sufficiently small, then we can use the proved lemma. Therefore, on S a generatrix will be found, which emanates from some point on the image $A_1 A_2$, consists of non-plane points and runs up to one of the points of the image of the segment $D_1 D_2$.

Consider now a plane domain in Figure 10. Let the domain with an l. E. metric be obtained by identifying the sides $A_1 B_1$ with $A_2 B_2$ and $K_1 N_1$ with $K_2 N_2$. Let the length of $A_1 A_2$ be equal to $2R$, the length of $A_2 B_2$ be h, the length of $C_2 D_2$ be equal to $s < h/2$, the length of $K_1 N_1$ be equal to $H > 2R$, the length of $K_1 K_2 \approx s$. Let this triply connected domain with the l. E. metric introduced with account for the indicated identifications and with the boundary $A_1 A_2 \cup K_1 K_2 \cup B_1 C_1 N_1 N_2 D_1 D_2 C_2 B_2 B_1$ be isometrically immersed into E^3 as some C^1-smooth torse S. Then, using the reasoning of the previous paragraph, we have that there are two generatrices l_1

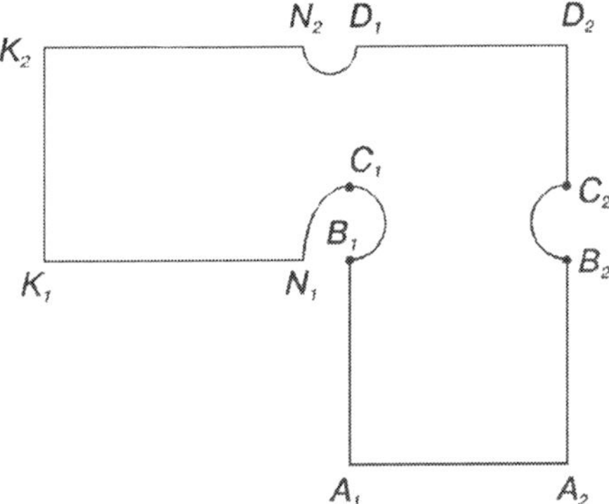

Figure 10

and l_2, of which one emanates from some point on the image of $A_1 A_2$ and runs to the intersection with the image of $D_1 D_2$, and the other emanates from some point on the image of $K_1 K_2$ and intersects with $D_2 C_2$. This means that these generatrices intersect each other, which contradicts their choice. This contradiction does prove that the considered domain does not admit an isometric immersion into E^3 even in a very weak class of regularity.

4.3   Isometric immersions of some particular domains with locally Euclidean metrics into R^3

It is known (see, for example, [36, 115, 192]) that two-dimensional manifolds carrying a complete l. E. metric are isometric to one of the following manifolds: the plane, cylinder, Möbius band, torus and Klein bottle (the respective domains on the plane with the known identification of the boundaries are referred to). It is natural to raise a question on the possibility of representing them as surfaces in E^3 or in spaces of higher dimension.

As for noncomplete l. E. metrics, they can be introduced on any manifold, open or with the boundary. Indeed, there are two inexhaustible sources to set them: these are polyhedra and minimal surfaces. Any polyhedron has a locally Euclidean metric everywhere except the vertices. Therefore, the deletion of all vertices of some polyhedron or an arbitrary neighbourhood of the vertices yields a manifold with an l. E. metric; if only the vertices or closed neighbourhoods of the vertices are deleted, we have l. E. metrics on open manifolds; but if some open neighbourhoods of the vertices are deleted, then we have l. E. metrics on manifolds with the boundary[25]. This construction of l. E. metrics is possible and abstract, as in [63] it has been proved that on any two-dimensional piecewise linear manifold we can introduce a polyhedral metric with *a priori* known values of curvatures in prescribed points inside the manifold (and at its boundary if any), certainly, with the required satisfaction of the Gauss–Bonnet theorem. Further, it is known that on minimal surfaces with curvature $K \leq 0$ and metric ds^2 the metric $ds_0^2 = \sqrt{-K}ds^2$ is locally Euclidean (this fact is known in the literature as the Ricci criterion for metrics of minimal surfaces, because the inverse proposition is also true: if for some metric ds^2 of Gaussian curvature $K < 0$ the metric $ds_0^2 = \sqrt{-K}ds^2$ is locally Euclidean, then ds^2 is the metric of some minimal surface, see, e.g., [39]). If $K \neq 0$, then the domain of definition of this l. E. metric has the same topological structure as the minimal surface itself; but if there are isolated points on a minimal surface, where $K = 0$, then we obtain l. E. metrics with conical points, therefore, such surfaces also generate polyhedral metrics. In [85], we can find the properties of such (and more generalized) metrics and their applications in the theory of minimal surfaces. There are also several other ways of introducing l. E. metrics and in general metrics of constant curvature on minimal surfaces, see about this, in particular, [84] and [85]. About other classes of surfaces, also generating constant-curvature metrics and, in particular, l. E. metrics, see [40].

Not only sets of manifolds with l. E. metrics but also sets of surfaces with flat metrics are abundant. It could be said that, in a sense, surfaces with flat metrics are "the same number" as surfaces without restriction to the type of metric. We have in mind the result of the work [133], in which the following propositions were proved: (a) in

---

[25] As an example of the relation of polyhedra to l. E. metrics, see recently published works [161] and [162], which investigate the problem of the isometric embeddings of polyhedra into E^3 in the form of piecewise smooth curved surfaces without any plane domain; [162], in particular, shows that all Plato polyhedra admit bendings (i.e., continuous isometric deformations) in the class of piecewise smooth surfaces.

a three-dimensional space, any compact surface with the non-empty boundary is isotopic to some surface with a flat metric; (b) any two compact surfaces with flat metrics are isotopic in the class of the same surfaces in that and only that case, when they are isotopic in the class of general surfaces.

4.3.1   On the isometric immersions of a complete locally Euclidean metric, given in some simply connected domain, into E^3

The following theorem can be proved.

Theorem 4.3. *Any complete l. E. metric (2.3) given in some open simply connected domain D can be isometrically embedded into E^3 as a plane. Herewith, if the metric is of class $C^{n,\alpha}$, then for the cases $n+\alpha > 0$ the embedding is given by an $C^{n+1,\alpha}$-smooth mapping, and if the given metric is continuous (i.e., $n+\alpha = 0$), then the embedding will locally be of class $C^{0,1}$ and there could be no class C^1.*

Indeed, in the domain D with an l. E. metric (and in general with any Riemannian metric) we can introduce a conformal structure, in which the metric in each chart has an isothermal form. The simply connected domain with a conformal structure is conformally equivalent in the large to either a circle or a plane, and the metric in it is analytical in isothermal coordinates. Under these conditions, the Killing–Hopf theorem ([190], Chapter 2) is already applicable, according to which a simply connected domain with the complete l. E. metric is isometric to the Euclidean plane. We will, however, adduce an independent and more illustrative proof of the existence of such an isometry in the large with its explicit construction. As we have already shown above in Remark 2.7, the simply connected domain Ω with an l. E. metric $ds^2 = \Lambda(u, v)(du^2 + dv^2)$ is immersed into the standard Euclidean plane R_{xy}[26]; herewith, this immersion is

---

[26] This immersion can be also obtained for the initial domain D; however, for the case of continuous metrics we would not be able to apply the following reasoning to geodesics, as in this case, first, more than one geodesic can emanate from one point in the given direction [79]; second, there can be points at which the emanating geodesic has no direction at the egress point, see [35] and [197]. But the proposition of the theorem is also valid for such metrics exactly due to the possibility of using the reasoning with their isothermal form.

performed by some holomorphic function $\Phi(w)$, $w = u + iv \in \Omega$, with $|\Phi'(w)| = \Lambda(w)$. Due to the local regularity of immersion, segments of straight lines in the pre-image are assigned smooth geodesic lines emanating from each point in Ω in a definite direction and uniquely defined by this direction. Owing to the completeness of the metric, each geodesic in Ω is extended infinitely, therefore, in R_{xy} it is assigned a complete straight line. Hence, from the single-valuedness of the mapping Φ it follows also that in Ω there could be no closed geodesic, otherwise its image would have been a closed geodesic on the plane. We fix in Ω some point M_0 and release geodesic rays from it in each direction. They will be assigned in R_{xy} rectilinear rays emanating from $\Phi(M_0) \in R_{xy}$, and in totality covering all the plane. Therefore, the mapping Φ is a surjection. Besides, this mapping is also injective. Indeed, let two points M_1 and M_2 from Ω be mapped into one point in R_{xy}. On the strength of the completeness of the metric, M_1 and M_2 are joined in Ω by a shortest geodesic, which on R_{xy} is assigned a rectilinear segment, so the images M_1 and M_2 cannot be coinciding points. This means that the immersion $\Phi(w)$ is in reality an embedding with $\Phi(\Omega) = R_{xy}$, which is what was proposed. And the smoothness of embedding $D \to R_{xy}^2 \subset R^3$ affirmed in the theorem is obtained from the theorem on the smoothness of isometries.

As the isometric embedding $\Omega \to R_{xy}$ is given by the conformal mapping $\Phi : \Omega \to R_{xy}$, the proof of the theorem yields the following corollaries.

Corollary 4.2. *If some simply connected domain D is a domain of definition of a complete l. E. metric in isothermal coordinates, it coincides with the entire plane.*

Indeed, for an isometric embedding of such a domain into the whole plane the inverse function will prove to be a single-valued analytical function defined for all the plane and conformally mapping all the plane to the domain D. If the domain D is bounded, this is impossible according to the Liouville theorem; but if D is not bounded, but does not coincide with the entire plane, then the existence of such a function is impossible according to the big Picard theorem.

Corollary 4.3. *Given for all the plane (x, y), the complete l. E. metric ds^2 in isothermal coordinates has the form $ds^2 = c(dx^2 + dy^2)$,* *where c is some positive constant.*

The next corollary shows that geometric interpretation may prove useful for producing some results in complex variable theory.

Corollary 4.4. *For any holomorphic function $w = f(z)$ given in a simply connected domain $\Omega \neq \mathbf{C}$ and not taking the value $w = 0$ there exist the routes γ_1 and γ_2 going to the boundary of the domain, along which, respectively, $\int_{\gamma_1} |f(z)||dz| < \infty$ and $\int_{\gamma_2} \frac{1}{|f(z)|}|dz| < \infty$.*

Indeed, in Ω we can introduce the l. E. metrics $ds_1^2 = |f(z)|^2|dz|^2$ and $ds_2^2 = \frac{1}{|f(z)|^2}|dz|^2$. As the simply connected domain Ω does not coincide with the entire plane, then both these metrics are not complete, whence we have the proposed. It is evident how the analogue of this proposition should be written down if the function $f(z)$ does not take some value $w = w_0$.

We complete this Section with the following:

Remark 4.8. Note that although the final isometric mappings $f : D \to R^2$ have only the finite smoothness and even of class $C^{0,1}$ only, their image – the plane – is an analytical surface (the smoothness of the mapping is lost in transition from the *a priori* given l. E. metric form to the isometric or standard form). This effect can be, certainly, for any metric, if the metric of a "good" surface is defined in "bad" coordinates and then this surface is considered to be obtained by isometric immersion of the metric form in these "bad" coordinates. This observation shows the possible difference between *the smoothness of the surface as the image* of an isometric immersion of a given metric ds^2, and *the smoothness of the immersing mapping* $\mathbf{r}$ *itself*, as a solution of the differential equation $d\mathbf{r}^2 = ds^2$: in the general case the smoothness of the surface as an image is not lower than the smoothness of the mapping defining it (this phenomenon was observed as early as by Yu.F. Borisov, see [197]). Correspondingly, the smoothness of the metric of a surface is not lower than the smoothness of that metric form by the isometric realization of which the surface was obtained. We recall that according to [150] the best "real" smoothness of the metric of a surface in Hölder classes $C^{n,\alpha}, n \geq 2, 0 < \alpha < 1,$

148   I.Kh. SABITOV

is achieved in harmonic coordinates, and in the two-dimensional case
in isothermal coordinates starting already from class $C^{1,\alpha}$. $\bullet$

4.3.2 Isometric immersions of complete locally Euclidean metrics, given in doubly connected domains, into E^3

Theorem 4.4. *Any complete l. E. metric (2.3) given in some doubly connected domain D can be isometrically embedded into E^3 in the form of a right circular cylinder.*

Indeed, by local transitions to isothermal coordinates a conformal structure can be introduced in D, which structure by the uniformization theorem will be conformal to some ring $\omega : 0 \leq \rho_1^2 < \xi^2 + \eta^2 < \rho_2^2 \leq \infty$ with a complete l. E. metric $ds^2 = \Lambda^2(\zeta)(d\xi^2 + d\eta^2), \zeta = \xi + i\eta$. By the Killing–Hopf theorem, this metric is isometric to that of a right circular cylinder of radius R with the metric

$$ds^2 = \frac{R^2}{|w|^2}(du^2 + dv^2), w = u + iv, \tag{4.9}$$

given in the domain $\omega_0 : 0 < |w| < \infty$.

Herewith, simultaneously we obtain that:

1) *the isothermal form of a complete l. E. metric on a doubly connected manifold can be given only on a sphere with two punctured points,* i.e., for the above mentioned ring ω we should have $\rho_1 = 0, \rho_2 = \infty$ (indeed, the isometry between the domains ω and ω_0 is given by the conformal mapping, but such a mapping is possible only at $\rho_1 = 0, \rho_2 = \infty$).

2) *the isothermal form of a given complete l. E. metric in the canonical domain ω_0 is unique* (indeed, for this it suffices to note that the conformal mapping of the domain ω_0 onto itself has the form $cw^n, n = \pm 1$ and leaves the metric form unchanged).

For functions holomorphic in the ring $0 < r_1 < |z| < r_2 \leq \infty$ or $0 < |z| < r_2 < \infty$, we can formulate analogues of the propositions of Corollary 4.3. It can even be admitted that the function $f(z)$ is many-valued, only then it is necessary to assume additionally the single-valuedness of $|f(z)|$.

The circular cylinder equation realizing the metric (4.9) is commonly known. But if "material" for the cylinder is given by a geometric description – as the band $\Pi(a, b)$ of the plane (ξ, η) between

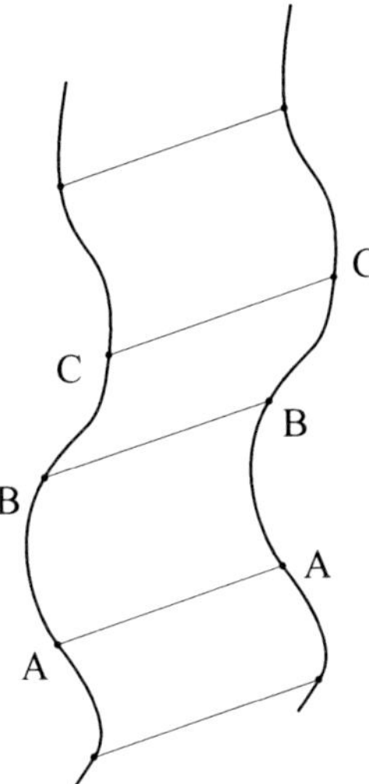

Figure 11

two non-intersecting curves $\xi = f(\eta)$ and $\xi = f(\eta - b) + a, a > 0, b > 0$ (see Figure 11), with identification of the points $(f(\eta), \eta)$ and $(f(\eta) + a, \eta + b)$, where $f(\eta)$ is a continuous curve defined for all $-\infty < \eta < +\infty$, and satisfying at $0 \leq t \leq 1$ the inequalities $f(\eta + tb) \leq f(\eta) + ta \leq f(\eta + tb - b) + a$ (they provide for the position of each straight segment between identified points to be inside the band $\Pi(a, b)$, and for them to be satisfied it is sufficient, e.g., to assume that $f \in C^1$ and $f'(\eta) \leq a/b$) – then the analytical isometric embedding of such a domain with the standard Euclidean metric into E^3 is given by a straight circular cylinder with the following position vector:

$$x = \frac{a}{2\pi \cos \alpha} \cos\left(\frac{2\pi \cos \alpha}{a}(\xi \cos \alpha + \eta \sin \alpha)\right)$$

$$y = \frac{a}{2\pi \cos \alpha} \sin\left(\frac{2\pi \cos \alpha}{a}(\xi \cos \alpha + \eta \sin \alpha)\right)$$

$$z = -\xi \sin \alpha + \eta \cos \alpha,$$

where $(\xi, \eta) \in \Pi(a, b)$, $\tan \alpha = b/a$. In particular, at $f(\eta) = c = const$, $b = 0$ the boundary lines $x = c$ and $x = c + a$ of the band pass into rectilinear generatrices, and at $b \neq 0$ the edges of the band (with the "skew" identification of the boundary points) pass into cylindrical helices. It is seen from the formulae that the isometric image from the strip Π into the standard band $0 \leq X \leq 2\pi R$ on the plane (X, Y) with the standard metric $dX^2 + dY^2$ is given by the formulae

$$X = R \, arg \, (x(\xi, \eta) + iy(\xi, \eta)), \quad Y = z(\xi, \eta), \quad R = \frac{a}{2\pi \cos \alpha} \quad \text{with the}$$

values of the function arg between 0 and 2π.

4.3.3   *Isometric immersion of a Möbius strip into E^3*

Every reader, probably, has glued a surface from a rectangular strip of paper or saw how it is glued, by such an "identification" of two short sides of a rectangle, at which its diagonals pass into closed lines. This glueing presents no big problem physically, on condition that the strip is long and narrow enough. A non-orientable surface would be obtained, which was discovered by Möbius back in 1865. But surprisingly enough the representation of at least one such surface through explicitly written out formulae proved not an easy problem, and it was solved only recently in [156]; however, in reality it is still not known if the simplest formula has already been found. Apparently, the papers [131, 157] and [38] contain the most essential bibliography on the subject, the literature cited in those works should only be supplemented with [70] (the list does not pretend to be complete, the more so that there are lots of text and graphic sites in the Internet on the Möbius band). We will call a non-orientable surface of zero curvature, embedded into E^3 and diffeomorphic to a rectangular Möbius strip (i.e., having, in particular, the one-component closed edge) a Möbius *band*, and if such a surface is *isometric* in the large to a rectangular Möbius strip, then it will be called a *standard* Möbius band. The curve on the Möbius band corresponding (at the isometry or diffeomorphism) to the midline of the rectangle, we will call the *midline* of the surface.

4.3.3.1 *Standard Möbius bands.*

The first work on the proof of the existence of an *analytical* standard Möbius band we are aware of was written by Wunderlich [193]. That work, however, contains no completely found formulae (though the midline in projective coordinates is written down explicitly through polynomial equations of degrees not higher than 6; about the surface itself, it could be only said that it is an algebraic surface of degree 39). In the non-analytical class, the standard Möbius band consisting of three plane and three cylindrical domains was constructed in $[152]^{27}$ (this construction can yield a smoothness of up to class C^∞). Chronologically, the next work on this subject appears to be [70], which proves that for the existence of an isometric *immersion*[28] of a rectangular Möbius strip into E^3 it is necessary and sufficient that the length L of the midline and the height h of the rectangle be related by the inequality $L > \dfrac{\pi}{2}h$. In [156] and [157], an example is constructed of the standard analytical Möbius band with high symmetry and, possibly, with the simplest midline, having the equation

$$x = \sin \varphi, \ y = (1 - \cos \varphi)^3, \ z = (1 - \cos \varphi)\sin \varphi, \ 0 \le \varphi \le 2\pi \quad (4.10)$$

and representable as the intersection of two algebraic surfaces of, respectively, degrees 4 and 6. But the explicit representation of the directions $\mathbf{l}$ of the generatrices of this surface as the functions of φ occupies too much space, although the writing down of the formula itself for $\mathbf{l}$ is rather simple

$$\mathbf{l} = sgn(\sin \frac{\varphi}{2})\mathbf{B} + \frac{\varkappa(\varphi)}{q(\varphi)}\mathbf{T},$$

where $\varkappa(\varphi)$ and $k(\varphi) = sgn(\sin \frac{\varphi}{2})q(\varphi) > 0$ are, respectively, torsion and curvature of the midline, $\mathbf{T}$ and $\mathbf{B}$ are unit vectors of the tangent and binormal of the midline. In [131], based on the idea of [156] with some assumptions for the class of C^n-smooth ($n \ge 3$) curves the necessary and sufficient conditions are given for the curve of this class to be a midline of some standard Möbius band. The work [146], also based on the idea of [156], proposes a criterion for midlines of analytical standard Möbius bands, in fact coinciding with the respective criterion from [131], but yielding a conceptually slightly different

---

[27] A nice picture of this band can be found in [157].

[28] Not an analytical immersion is meant *a fortiori*.

description of the class of admissible curves. In what follows we adhere to [146].

Thus, let $\mathbf{r}(s) = \{x(s), y(s), z(s)\}, 0 \le s \le L$, be a natural equation of the midline Γ of a standard Möbius band S, and L be the length of the midline, which, as the image of a straightline segment, should be a geodesic on S. In this case, the principal normal $\mathbf{N}$ to the midline, being parallel to the normal to the surface, should after walking along the curve arrive at the opposite position, i.e., it is antiperiodic: $\mathbf{N}(s + L) = -\mathbf{N}(s)$. But $\mathbf{r}_{ss} = k_A(s)\mathbf{N}(s)$, so the analytical curvature k_A introduced in Subsection 3.2.2 is also an antiperiodic function, therefore, it should have an odd number of zeros on Γ, with account for multiplicity. And from the same Subsection 3.2.2 it is known that in curvature zeros the torsion of the curve should have zeros of no smaller order than the curvature[29]. If this condition is satisfied, then the sought-for standard Möbius band does exist, and its position vector is given by the formula (3.17). Thus, we proved the following:

Theorem 4.5. *For the analytical closed curve Γ to be the midline of an analytical standard Möbius band, it is necessary and sufficient that its curvature be antiperiodic and that in points of curvature inversion to zero the torsion of the curve have zero of no smaller order than the curvature. Herewith, the Möbius band even in class of C^2-smooth surfaces is determined by its given analytical midline uniquely in the form of an analytical surface with the equation*

$$S : \mathbf{r}(s, t) = \{x(s), y(s), z(s)\} + t(\cos \alpha \mathbf{T} + \sin \alpha \mathbf{B}), \qquad (4.11)$$

in which

$$\sin \alpha = \frac{1}{\sqrt{1 + h^2(s)}}, \quad \cos \alpha = \frac{h(s)}{\sqrt{1 + h^2(s)}}, \qquad (4.12)$$

---

[29] Besides, it should be noted that the torsion of *any* closed analytical curve is a periodic function, as the known classical expression for torsion involves only derivatives of the position vector of the curves, and in curvature zeros this expression admits the extension of the definition by continuity. That is why in classical works it is usually assumed that the curvature $k \ne 0$, otherwise without analyticity in points where $k = 0$, it is impossible to determine torsion.

where the function

$$h(s) = \frac{\varkappa(s)}{k_A(s)}$$

is considered by definition to be extended by continuity in curvature zeros.

This theorem and its proof can be used to obtain various geometrically and analytically useful facts, which we will represent as a number of corollaries and remarks.

Corollary 4.5. *On a standard analytical Möbius band the difference between the number of zeros (with account for multiplicity) of torsion of the midline and its curvature is an odd number (and, therefore, there is at least one point, where the order of the zero of torsion is greater than the order of curvature zero[30]).*

Indeed, we know that the torsion of a closed analytical curve is periodic, and for the midline of the standard analytical Möbius band the analytical function $h = \varkappa(s)/k_A(s)$ is antiperiodic, so the number of its zeros is odd.

Remark 4.9. In equation (4.11), we can substitute the variable

$$\xi = s + t\cos\alpha, \ \eta = t\sin\alpha, \tag{4.13}$$

which will lead the metric of the surface to the standard form $d\xi^2 + d\eta^2$. Herewith, as it should be, the midline passes into the segment $\eta = 0, 0 \leq \xi \leq L$, and the generatrix at the point $s = 0$ passes into the straight lines $\xi = 0$ and $\xi = L$ (recall that our option was $\sin\alpha(0) = 1$). If the equalities (4.13) are inversible (theoretically this is guaranteed in the neighbourhood of the segment $\eta = 0$), then we obtain explicit formulae for the isometric immersion of a rectangular Möbius strip on the plane (ξ, η) into R^3. Further, let a rectangular Möbius strip have a restriction $|\eta| \leq C$. Then due to (4.12) and (4.13) we obtain the satisfaction of the inequality

$$C < min\frac{1}{|h'(s)|}, \ h = \frac{\varkappa(s)}{k_A(s)},$$

---

[30] Included here are, certainly, those possible points, where torsion is equal to zero and curvature is different from zero.

which, together with the known length L of the base of the rectangle, is what yields the necessary restriction for the height $2C$ of the rectangle. Therefore, there is hope for a solution of the problem posed in [70] on the relation of the length and height of the Möbius rectangle admitting an isometric embedding into R^3.•

Remark 4.10. Note that in classes of finite smoothness, when the curvature passes through zero of finite order, the curves also admit the introduction of a smooth Frenet trihedron on them, in the assumption that the second derivative of the position vector of the curve has a local representation with the isolation of the main term, admitting the differentiation of the equivalency. Under these conditions, all the above adduced considerations are also possible for non-analytical curves.•

Remark 4.11. Since in continuous deformation of analytical curves (in the same classes of smoothness) the property of the principal normal – and, respectively, of the binormal – to be periodic or antiperiodic does not change, by the same token it turns out that *all closed spatial analytical curves are split into two non-intersecting classes – with periodic or antiperiodic principal normals and binormals entering the analytically smooth field of the Frenet trihedron.* Closed curves with the antiperiodic principal normal and binormal shall be further called *semi-periodic.*•

Remark 4.12. A useful observation for some problems is that on the plane and on convex surfaces there are no closed curves with antiperiodic normals. Indeed, on the plane the normal to the curve can be defined as a periodic vector obtained by the turn around $90°$ of the vector of the tangent, and on the convex surface no curve of smoothness C^2 can have an alternating-sign curvature, which is required for the principal normal to be antiperiodic.•

Open question 4.2. Since closed spatial curves, at least in the analytical class, are split into two non-intersecting classes, it could turn out that in the theory of knots there is sense to consider separately deformations with preservation of the class of curves. In [132], this is done not for knots of curves but for knots tied on lying and narrow oriented or non-oriented bands glued from rectangles of respective form. Besides, it can be of interest to introduce some metric requirements for deformed curves – to impose some restrictions for their lengths, curvatures, torsions, positions on a surface, etc.•

4.3.3.2 Unsolved problems. Turning back to the issue of the "simplest" representation of an analytical standard Möbius band, we note the following approaches possible here. Firstly, it could be the simplest (in a sense) form of asymptotic parametrization of the band

$$\mathbf{r} = \mathbf{a}(u) + v\mathbf{b}(u), \qquad (4.14)$$

where u is some parameter for the description of the midline of the band (e.g., u is a natural parameter or a standard variable changing from 0 up to 2π for periodic trigonometric functions); $\mathbf{b}(u)$ is a unit vector of the generatrix of length $2v$, with some boundary of the change of $c_1(u) \leq v \leq c_2(u)$, corresponding to the size of the immersed rectangular Möbius strip (for example, we can start with the search for the lowest order for representing $\mathbf{a}(u)$ of the directrix through the trigonometric polynomials of the parameter φ, the way it is done in [156] and [131]). Secondly, we can talk of the "simplest" form of the equation for that algebraic surface, on which the constructed standard Möbius band is located (e.g., as we already said earlier, the surface described in [193] has in projective coordinates the degree 39, and for the standard Möbius band found in [156] it is only known to be located on some algebraic surface, but the explicit algebraic description is given only for the midline written down as the intersection of two algebraic surfaces of degrees 4 and 6)[31]. Thirdly, works by Sadovski ([152] and [153]) propose one variational problem the solution of which should be some standard Möbius band. Everybody who glued a Möbius band, probably, noticed that as the edges of a rectangular strip are brought to their required coincidence in space, the strip as if itself takes the shape of some surface without any "force action" by the experimenter; what is more, if we try to change the shape of the surface slightly by pressing some of its parts and then withdrawing the pressure, the strip returns to its initial

---

[31] In a personal communication to the author, Yu.G. Nikonorov reported that, by the computations of his student S. Poletayev, the Möbius band constructed in [156] is located on an algebraic surface of degree 20.

shape. This means that there are some internal stresses in the material, which automatically give the sheet of paper some optimal shape, thus minimizing these stresses. Sadovski found a variational problem corresponding to his interpretation of these stresses as elasticity forces distributed along the area inversely proportional to the square of the radius of curvature of the surface. This leads to the problem of the minimization of the integral $\int\int H^2 dS$, where H is the mean curvature of the surface (Sadovski himself proposed in [153] some one-dimensional variational problem, which is also described in [157]). Well then, the problem of the "simplest" form of the standard Möbius band can be formulated also as the problem of finding an equation of the surface minimizing the written down functional in a reasonable way, and from this point of view this problem still remains open.

In recent years, Sadovski's ideas have been applied and developed in works by E.L. Starostin and G.H.M. van der Heijden [168] and [169], which theoretically and numerically study the form of a flat Möbius band it takes under the influence of internal elasticity forces. This form turns out to depend significantly on the width of the band, and the characteristic property in the distribution of the generatrices is that a point of intersection of the generatrices emerges at the edge of the band, i.e., at the same midline the band cannot become wider or narrower with equilibrium preserved without changing the structure of the generatrices.

4.3.3.3 Some general propositions on flat Möbius bands. Before we pass to the construction of general Möbius bands, let us prove the following:

Theorem 4.6. *In an analytical class of immersions (i.e., also without the requirement of the absence of self-intersections) flat Möbius bands can be only torses, i.e., developments of the tangents to some spatial curve.*

Let us prove this theorem. Indeed, let S be some analytical Möbius band. As the Möbius band is by definition diffeomorphic to a rectangular Möbius strip, we can consider on S a closed curve Γ, being the image of the midline of the Möbius strip. The curve Γ will also be called the midline of the surface S. The surface is analytical and, therefore, contains no plane domains, which means that through each its interior point, in particular, through each point of the curve Γ, a unique rectilinear generatrix passes, which runs to both sides up to the edge of the surface. Let the surface S be cylindrical. We will show that this contradicts its non-orientability.

All generatrices of a cylindrical surface are parallel to one direction l. We will consider that this direction is parallel to the axis z, i.e., $l = \{0, 0, 1\}$. For greater clarity, we will assume that all the surface is located above the plane (xy). Let $\mathbf{r}(s)$ be the position vector of the midline Γ in the natural parameter. The normal to S is determined by the direction of the vector $[\mathbf{r}'(s), l]$, and if no generatrix is tangent to the midline, then after its walking along, the normal would return to the initial position. Therefore, on Γ there is a finite number of points, at which $\mathbf{r}'(s) \parallel l$. We cover the midline with a finite number of open neighbourhoods U_i, in the interior of each of which S is a cylindrical surface, with the boundary consisting of two segments of the generatrix and two sections of S by the horizontal planes $z = C_i'$ and $z = C_i''$. Besides, we will assume that part Γ_i of the curve Γ within the limits of U_i is a simple Jordan arc with possible contingency with generatrices not more than at one point. The sections $z = C_i'$ and $z = C_i''$ within the limits of U_i are projected on the plane (xy) into the same regular analytical curve γ_i. The normal to the cylindrical surface with the plane directrix $\gamma_i : (x_i(t), y_i(t), 0)$ is determined by the choice of one of the normals to γ_i, say, the normal $\mathbf{N} = (-y'(t), x'(t), 0)$. If we knew that the combination of the curves γ_i represented a regular closed curve, then from the formula for the normal $\mathbf{N}$ we would have immediately obtained the required contradiction with the non-orientability of the surface S. But in principle this combination can fail to be the immersion of the circumference[32], so more thorough analysis is required to establish the orientability of S.

Let the number of neighbourhoods U_i covering Γ be equal to n, and let them be chosen such that the arcs $\Gamma_i \subset \Gamma$ lying in them consecutively continue one another when intersecting with their neighbours along small arcs. Let $A_1, A_2, \ldots, A_n$ be the initial points of

---

[32] And the projection of the midline, though it is a closed curve, in the tangency points with the generatrices will have singular points, in which the curve goes "back" along itself.

these arcs (at some fixed direction of the walk along the midline), and $B_1, B_2, \ldots, B_n$ be their endpoints, so that they can be considered to be ordered according to the rule conditionally written in the form of inequalities $A_i < B_{i-1} < A_{i+1} < B_i < A_{i+2} < B_{i+1}$ etc. As the tangency points of the midline Γ with the generatrices are isolated, the arc Γ_i in the neighbourhood of the tangency point is either at one side of the generatrix (we will call this point the *tangency point of the first kind*) or it passes into the tangency point from one side of the generatrix to the other side (similar to the plane curve in the neighbourhood of the point of its inflection). In the second case, the projection of the arc Γ_i coincides with γ_i – the projection of the section U_i by the horizontal plane, so the normal to S along Γ_i will coincide with the analytically changing normal $\mathbf{N}$ to γ_i. As before and after the passage through such a tangency point the direction of the normal to S is expressed by the same formula $[\mathbf{r}'(s), \mathbf{l}]$ and, in the final analysis, does not affect the periodicity of the normal, Γ should without fail have tangency points of the first kind on it. During the passage through these points the expression for the direction of the normal changes its sign (herewith the normal itself changes continuously, remaining constant along the generatrix). In this case, the projection of the arc Γ_i also remains on the curve γ_i, but this time it covers twice some of its part located within the limits of the projection U_i up to the point whereinto the generatrix on S tangent to the arc Γ_i is projected, and does not project at all into the other part of γ_i. Herewith, the normal to S along Γ remains the normal to the cylindrical surface, preserving the continuity in passage through the tangency point Γ_i with the generatrix, so it coincides with the previous normal $\mathbf{N}$ to γ_i in respective points[33]. During its movement along the arc Γ_i its projection passes twice along the respective part of the curve γ_i, being analytically extended along that curve γ_{i-1} or γ_{i+1}, which corresponds to the passage of a point on S into the neighbourhood U_{i-1} or U_{i+1}.

Let the projections of the points A_i to the plane (xy) be designated as A_i', and of the points B_i as B_i', $1 \leq i \leq n$; and as the arc Γ_1

---

[33] Formally, $\mathbf{N}$ should have been written down now as $(y'(t), -x'(t), 0)$, bearing in mind that the tangent to γ_i during the reverse walking along γ_i changes its direction to the opposite, but for brevity of the accompanying record we will limit ourselves by the verbal expression "previous normal".

let that arc on Γ be chosen, on which there is no tangency point with the generatrix, and the extension in the form of a straight generatrix at the point A_1 be nowhere tangent to the line Γ, either (this can always be achieved by decreasing the size of the neighbourhood U_1 and increasing the number of neighbourhoods). For greater clarity in our considerations, we can also assume that the point A_1 is the lowest point at the straight line, whose segment(s) serve(s) as generatrices to S along the midline Γ.

Then all the arc Γ_1 will project onto the arc γ_1 and the passage along Γ_1 is assigned the passage along γ_1 from the point A_1' up to the point B_1'. Further, the projections of the arcs $\Gamma_2, \ldots$ coincide with $\gamma_2, \ldots$ till the first arc $\Gamma_j, j \geq 2$, at which for the generatrices there is some tangency point $C_1 \in \Gamma_j = \smile A_j B_j \subset \Gamma$ of the first kind with the line Γ (because at least one such point does exist). Then the movement along Γ from C_1 up to the point B_j (and also, possibly, farther) is assigned the inverse passage along γ_j (and, possibly, also farther); what is more, the normal to S will coincide with the previous normal $\mathbf{N}$ to γ_j. Let this inverse passage along the projection Γ at the line Γ itself be assigned some arc $C_1 C_2$ along the direction of the walk along the midline (where C_2 is yet another point at which the generatrix is tangent to the midline, if, certainly, such a point, besides C_1, does exist), and in the projection – the route from C_1' up to C_2', with the straightforward notation of the projections of the points C_1 and C_2, respectively (Figure 12).

Assume first that in the projection the inverse passage did not reach the point A_1'. Then in the projection we will eliminate the route $C_2' C_1' C_2'$ passed twice in opposite directions, and will continue the walk along Γ from the point C_2; herewith, the projection of the line Γ will first again go along the previous projection $C_2' C_1'$ with the coincidence of the normal to S with the normal to the projection. Further, we repeat the previous construction: each time we eliminate the twice-passed segment of the projection (if the return does not take us to the initial point A_1'). As the result, we will have in the projection a once-passed closed regular analytical curve with the origin and end at the point A_1' (possibly, with self-intersections), the normal $\mathbf{N}$ to which will prove periodic, which is what will lead to the sought-for contradiction with the non-orientability of the surface S.

Let it turn out that in the previous considerations at some inverse

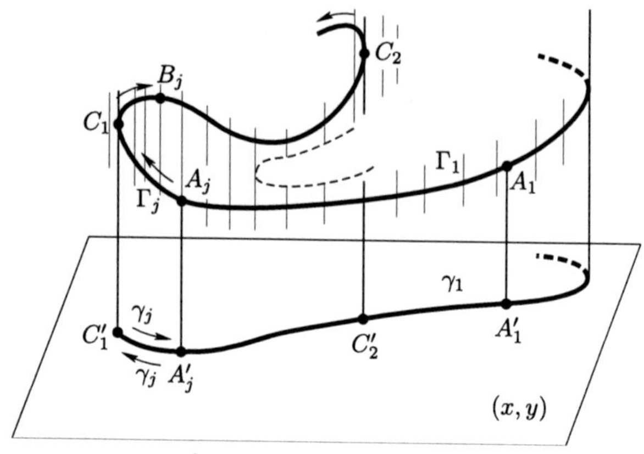

Figure 12 (courtesy of *Izvestiya: Mathematics*)

pass we arrive in the projection to the initial point A'_1. On the line Γ we mark the point C_*, at which this happened for the first time (in Figure 12 the point C_* is not shown). Of the projection of all the site $\smile A_1 C_* \subset \Gamma$, only one point will remain, which is the common projection $A'_1 = C'_*$ of the points $A_1 \in \Gamma$ and $C_* \in \Gamma$. At these points, the surface S has similarly directed normals (namely, the normal $\mathbf{N}$ to the curve γ_1 at the point A'_1). The projections of the surface S in small neighbourhoods of the points A_1 and C_* coincide and contain a common arc with the curve γ_1 with the common direction of the normal $\mathbf{N}$. We repeat the previous constructions, continuing the walk along Γ from the point C_* in the earlier chosen direction. And either we will pass the remaining part of the regular curve Γ, returning in the projection into the point $A'_1 = C'_*$ by the closed regular curve with the normals coinciding at the beginning and end of the route (it could also be that the curve γ after cancelling twice passed segments will degenerate into the point A'_1, preserving in the limit at the end of deformation that direction of the normal, which was in A'_1 at the beginning) or else we will come to some point $C_*^{(1)} \in \Gamma$ with the complete cancelling of twice passed segments, but with the coincidence of the directions of the normals to S at the points A_1, C_* and $C_*^{(1)}$. As a result, repeating this process several times, we will make sure that the normals to S along Γ at the beginning and end of the walk coincide, which was required to obtain a contradiction with the non-orientability of S.

Assume now that the Möbius band S is a conical surface. Then we will consider intersections of small domains of the surface S with a sphere of sufficiently large radius and with the centre at the vertex of the cone; and by reasoning similar to previous considerations will arrive at the contradiction with the non-orientability of S. Due to the analyticity, the variant remaining for S is only a torse.

Remark 4.13. As the expression for the direction of the normal, given by the vector product $[\mathbf{r}'(s), \mathbf{l}]$, in transition through a tangency point of the first kind formed with the midline changes for that opposite in sign, the non-orientability of the cylindrical Möbius band is equal to the oddity of the number of the tangency points of the first kind. For this reason, another proof of the impossibility of the existence of an analytical Möbius band in the form of a cylindrical surface could be obtained by establishing the parity of the number of the tangency points of the first kind.•

Remark 4.14. At the very beginning of the proof, it was established that if the generatrices are nowhere tangent to the midline, then the theorem is true. It is important to note that this is also valid for the class of surfaces of smoothness $C^n, 2 \leq n \leq \infty$.•

Remark 4.15. The main point in the proof of the impossibility of the existence of an analytical Möbius band in the form of a cylinder was the uniqueness of extending the projection of the surface in the neighbourhood of each generatrix irrespective of the position of the projected domain along the generatrix, on condition of the coincidence of projections on an arbitrarily small arc. Therefore, the theorem also remains valid in the classes of smoothness $C^n, 2 \leq n \leq \infty$, and in general for C^1-smooth cylinders on condition of the passage of a unique generatrix through each point, if we assume the uniqueness of extending the projections as in the case of analytical surfaces. This is the case on algebraic cylinders with the equation of the form $F(x, y) = 0$, which in the neighbourhood of the points $grad\ F(x, y) = 0$ may not have the analytical representation of the form $y = f(x)$ or $x = g(y)$.

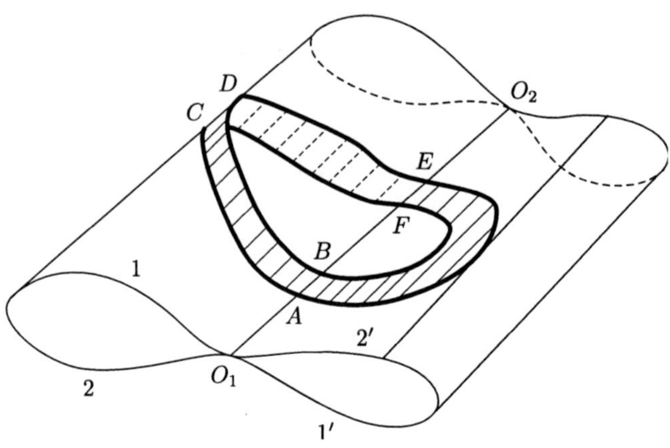

Figure 13 (courtesy of *Izvestiya: Mathematics*)

For instance, this is the case on the surface of a cylinder with the equation $x^6 - x^3(y^5 + y^7) + y^{12} = (x^3 - y^5)(x^3 - y^7) = 0)$; the uniqueness of extending the projection means here that over the entire extension of the cylinder in the neighbourhood of the axis z, being a double generatrix passing through the point $(0, 0, 0)$, the projection of the cylinder over the curve, e.g., $x^3 = y^5$, $x > 0$, is extended to the same curve – either to the curve $x^3 = y^5$, $x < 0$ (and then we will have an algebraic analytical cylindrical surface) or else to the curve $x^3 = y^7$, $x < 0$ (then a $C^{1,2/3}$-smooth cylinder will be obtained). The condition of analyticity makes the choice of extension unique, but if we will require from a Möbius band on this cylinder the uniqueness of extending its projection on the plane (x, y) over the entire extension of each generatrix as a straight line, then it will not be able to settle on the cylinder.$\bullet$

Remark 4.16. The surface shown in Figure 13 (constructed by the idea of the respective figure from [38]) shows that without the condition of analyticity or without the condition of uniqueness of extending the projections, described in Remark 4.8, embedded Möbius bands being cylindrical surfaces can exist even in class C^∞.

In this figure, the Möbius band consists of the part on a cylinder over curve 1, which is then extended to the cylinder over curve 2, and further to the cylinder over curve 2'. All of the cylinder can be considered to be analytical, assuming the analyticity of curves $1 + 1'$ and $2 + 2'$; in this case, the contingency between the curves along the generatrix $O_1 O_2$ will be of a finite order and the respective cylindrical (hatched) embedded Möbius band will have the smoothness of a finite order. But if at least one of these curves will be of class C^∞, the contingency between the curves can be of any order $n, 1 \leq n \leq \infty$, and the Möbius band will be of the same class of smoothness. The explanation of the possibility of such a situation is clear – along the generatrix $O_1 O_2$ the projections of the surface in the neighbourhood of the segments AB and FE do not coincide, though they do have the common arc on curve $2'$, i.e., the extension of the projection of the Möbius band from the site on curve $2'$, then to the left from the point O_1 is not unique. Note that in this surface the midline has only one (i.e., an odd number) tangency point of type 1, in the neighbourhood of which the surface is analytical, therefore, the proof of the theorem by the idea of Remark 4.6 does not seem to be an easy task.•

Remark 4.17. The cylindrical Möbius band in the given example is not the standard Möbius band. This is not by chance, as the standard Möbius band S cannot be a cylindrical surface even in the class of C^1-smooth torses in the sense of Chapter 3. Indeed, the midline of the standard Möbius band is on S a smooth closed geodesic, nowhere tangent to its generatrices of constant direction $\mathbf{l}$, so the normal $[\mathbf{r}'(s), \mathbf{l}]$ to S will return after walking along the midline to the initial position.•

From this remark, it follows that a *metrically complete Möbius band could not be immersed into E^3 even in the form of a C^2-smooth surface* and, more generally, even in the form of C^1-smooth torses in the sense of Chapter 3. Indeed, if such a surface S does exist, then its double cover will be a complete oriented developable surface, but which, according to the Pogorelov theorem and its known (from Chapter 3) generalization for C^1-smooth torses, is a cylinder. In this case, the initial surface S, too, is a cylinder, and we can single out on S a domain representing a cylindrical standard Möbius band around the midline, which is impossible.

4.3.3.4 Construction of general flat Möbius bands. The general procedure of constructing a flat Möbius band in E^3 is obtained from the following considerations. We will search for an analytical developable surface in asymptotic parametrization (4.14) in the assumption that the generatrices are nowhere tangent to the directrix. Let the directrix be given as an analytical[34] midline Γ with 2π-periodic position vector $\mathbf{r}(\varphi) = \{x(\varphi), y(\varphi), z(\varphi)\}, 0 \leq \varphi \leq 2\pi$ (the parameter φ we do not yet consider natural, assuming that we search for a Möbius band with the directrix midline of arbitrary length). Let $\mathbf{l}(\varphi) = \{a(\varphi), b(\varphi), c(\varphi)\}$ be a nonzero vector of the direction of the generatrices, nowhere tangent to the directrix. Then on the strength of Proposition 3.5 the respective asymptotic parametrization of the analytical Möbius band will be analytical; consequently, all functions involved in further considerations can be considered to be analytical. The direction of the normal $\mathbf{n}$ to the sought-for surface S

$$x = x(\varphi) + ta(\varphi), y = y(\varphi) + tb(\varphi), z = z(\varphi) + tc(\varphi) \qquad (4.15)$$

is determined by the vector product $[\mathbf{r}'(\varphi), \mathbf{l}(\varphi)]$, and after the walk along the directrix it should change to the opposite. As the vector $\mathbf{r}'(\varphi)$ tangential to the curve Γ is periodic, then the change of direction of the normal is possible only at the expense of the change of direction of the vector of the generatrix. Therefore, we have the conditions $a(\varphi + 2\pi) = -a(\varphi), b(\varphi + 2\pi) = -b(\varphi), c(\varphi + 2\pi) = -c(\varphi)$, i.e., the components of the vector $\mathbf{l}$ should be 2π-*antiperiodic*. This means that each component of the vector $\mathbf{l}$ somewhere turns to zero. Let us use this fact to prove the following:

Proposition 4.1. *There exists no analytical Möbius band with the plane midline.*

We will begin the proof of this proposition with the following lemma.

Lemma 4.4. *On an analytical developable surface distinct from a plane domain, the tangent plane locally intersects with the surface only by the generatrix passing through the tangency point.*

Indeed, let the point $(0, 0, 0)$ be a tangency point and $z = 0$, the tangent plane. Let the axis Ox be directed along the generatrix and

---

[34] On any analytical Möbius band it is easy to construct an analytical curve, which can be taken for the midline of the band.

let $\mathbf{r}(s) = \{0, y(s), z(s)\}, -\varepsilon < s < \varepsilon$, be the analytical position vector of the directrix on the plane $x = 0$ and $\{a(s), b(s), c(s)\}$ be an analytical unit vector of the generatrix. Then for the surface we have the following equation:

$$x = ta(s), y = y(s) + tb(s), z = z(s) + tc(s).$$

By the conditions for the arrangement of the surface we have the following representations:

$$y(s) = y_0 s + ..., y_0 \neq 0, z(s) = z_0 s^n + ..., z_0 \neq 0, n \geq 2,$$
$$a(s)=1+a_0 s^m+..., a_0 \neq 0, m \geq 1, b(s)=b_0 s^p+..., b_0 \neq 0, p \geq 1, (4.16)$$
$$c(s) = c_0 s^q + ..., c_0 \neq 0, q \geq 1.$$

Besides, we have a condition for the developmentability of the surface

$$(ca' - ac')y' + (ab' - ba')z' = 0.$$

From this condition with account for the relations (4.16) we have that $q = n + p - 1$. But this equality, when applied to the equation $z = 0$, yields that it has only one solution $s = 0$, i.e., the generatrix $(t, 0, 0)$. The lemma is proved.

Let us proceed to prove the proposition. First of all, note that, as the surface is analytical, any plane line on it, as the intersection with the plane, is analytical, therefore the assumed plane midline Γ on the considered band S should be analytical. Let the plane of the curve Γ coincide with the plane $z = 0$ and let $\mathbf{r}(\varphi) = \{x(\varphi), y(\varphi), 0\}$ be its position vector, and $\mathbf{l}(\varphi) = \{a(\varphi), b(\varphi), c(\varphi)\}$ be a unit vector of the direction of the generatrices. First we assume that the generatrices are nowhere tangent to the midline. Then in a sufficiently narrow band along Γ the surface has an analytical representation (4.15), and the unit normal $\mathbf{n}$ to the surface along the entire curve Γ can be calculated from the vector product $\mathbf{n} = [\mathbf{r}'(\varphi), \mathbf{l}(\varphi)]$. We have:

$$\mathbf{n} = \frac{1}{\sqrt{(bx' - ay')^2 + c^2(x'^2 + y'^2)}}\{cy', -cx', bx' - ay'\} . \qquad (4.17)$$

The condition of stationarity of the normal to S along the generatrices is given by the equality of the mixed product $(\mathbf{r}', \mathbf{l}, \mathbf{l}')$ to zero, which leads to the equation

$$c'(bx' - ay') + c(a'y' - b'x') = 0. \qquad (4.18)$$

The functions a, b, c are antiperiodic, therefore on Γ there is a point $\varphi = \varphi_0$, at which $c(\varphi_0) = 0$. From the equation (4.18) we have that at this point there should be $bx' - ay' = 0$, as in the analytical function c zero has the order higher than the possible zero in its derivative c'. But then it will prove that at this point the vector $\mathbf{l}$ is tangent to the midline, which is impossible by the assumption.

Consider now the case when on the line Γ there is one or several points, at which it is tangent to the generatrices. In this case we cannot represent the surface in the large as (4.15), but it is easy to show that along Γ the directions of the generatrices $\mathbf{l}$ all the same have an analytical dependence on φ, as the locally analytical representation of the form (4.15) remains valid with some other analytical directrix not tangent to the line Γ. Then in transition through the tangency point the formula (4.17) for the normal can change its sign for the opposite sign, i.e., the continuously changing direction of the normal to S can be determined on Γ on different sides from the tangency point by vector products $[\mathbf{r}', \mathbf{l}]$ and $[\mathbf{l}, \mathbf{r}'] = -[\mathbf{r}', \mathbf{l}]$. As the direction $\mathbf{l}$ is antiperiodic, to obtain the opposite direction of the normal at the end of the walk along Γ we should arrive to the initial choice of the sign of the product $[\mathbf{r}', \mathbf{l}]$. This means that the number of changes of the sign before the vector product should be even. Where the generatrix is tangent to the line Γ, we should have $c = 0$, $bx' - ay' = 0$. Let the order of zero of $c(\varphi)$ at the point $\varphi = \varphi_0$ be equal to n, and for the function $h(\varphi) = bx' - ay'$ it is equal to m. Let $n > m$. Then, passing in (4.17) to the limit at $\varphi \to \varphi_0$ we see that the tangent plane to the surface at the point $\varphi = \varphi_0$ coincides with the plane of the midline. This means that the tangent plane intersects the surface by at least two lines – by the generatrix $\mathbf{l}(\varphi_0)$ and by the curve Γ, which is impossible by Lemma 4.1. Therefore, always $n \leq m$.

Let $n = m = 2p$. Then $c(\varphi) = c_0(\varphi - \varphi_0)^{2p} + ..., c_0 \neq 0, h(\varphi) = h_0(\varphi - \varphi_0)^{2p} + ...$ and the formula (4.17) for the calculation of the normal takes the form

$$\mathbf{n} = \frac{s^{2p}}{|s|^{2p}\sqrt{A_0^2 + ...}}\{c_0 y' + ..., -c_0 x' + ..., h_0 + ...\}, \quad s = \varphi - \varphi_0,$$

where $A_0 \neq 0$. This means that in this case during the passage of the function $C(\varphi)$ through zero the requirement of continuity of the normal $\mathbf{n}$ leads to the necessity of preserving its expression $[\mathbf{r}', \mathbf{l}]$ both from the left and from the right of φ_0. But if $n = m = 2p+1$, then to preserve continuity of the normal the formula should change its sign.

Let now $n < m$. Then the sign in the formula changes again only at odd n. As the number of changes of the sign should be even, as the result we obtain that the total number of zeros of the antiperiodic function $c(\varphi)$ with account for multiplicity should be even, which is impossible. The proposition is proved.

Remark 4.18. Lemma 4.1 proved for an analytical case is already not true in classes $C^n, 2 \leq n \leq \infty$, but it is true at an additional assumption that the curvature of the plane directrix is different from zero, i.e., when the considered point of the surface is not a point of flattening. Therefore, our proof of Proposition 4.1 also covers the cases of non-analytical Möbius bands at additional assumptions: 1) the curvature of the plane midline is everywhere different from zero (i.e., this line is strictly convex); 2) possible points of tangency of the midline with the generatrix are isolated; 3) the tangency in them is of integer order. This gives us another proof of Proposition 4.1 in conditions under which a similar proposition was proved in [38].$\bullet$

Let us continue with the deduction of the equation of a developable Möbius band in the general case. From the condition of constancy of the normal along the generatrices, we obtain the equation

$$(\mathbf{r}', \mathbf{l}, \mathbf{l}') = (cb' - bc')x' + (ac' - ca')y' + (ba' - ab')z' = 0. \quad (4.19)$$

Therefore, to construct an embedded Möbius band, it is sufficient to find three 2π-periodic functions x, y, z and three 2π-antiperiodic functions a, b, c satisfying the equation (4.19) and three conditions: 1) $a^2 + b^2 + c^2 \neq 0$; 2) the spatial curve $\Gamma : \mathbf{r}(u) = \{x(u), y(u), z(u)\}$ has no self-intersections; 3) the vector $\{a, b, c\}$ is nowhere tangent to the curve Γ (the first two conditions are necessary). Then the equations (4.15) at sufficiently small $|t|$ define the surface, containing as its part an isometric image of some quadrangular (in the general case, curvilinear) domain on the plane with identification of one pair of the opposite sides of equal length with their inversion in identification (see Figure 16).

By the assumption, no generatrix is tangent to the directrix, so the vectors $\mathbf{r}'$ and $\mathbf{l}$ are linearly independent. Therefore, the vector $\mathbf{l}'$ is linearly expressed through these two vectors: $\mathbf{l}' = \mu\mathbf{l} + \nu\mathbf{r}'$, where $\mu(\varphi)$ is a 2π-periodic and $\nu(\varphi)$ a 2π-antiperiodic function. This relation leads to an inhomogeneous linear differential system

$$a'(\varphi) = \mu a + \nu x', \quad b'(\varphi) = \mu b + \nu y', \quad c'(\varphi) = \mu c + \nu z'$$

or, in an abbreviated vector form,

$$\mathbf{l}'(\varphi) = \mu(\varphi)\mathbf{l}(\varphi) + \nu(\varphi)\mathbf{r}'(\varphi), \tag{4.20}$$

at the arbitrarily chosen 2π-periodic function $\mu(\varphi)$ and 2π-antiperiodic function $\nu(\varphi)$ with the condition $[\mathbf{r}', \mathbf{l}] \neq 0$. The same equation is also suitable in the non-analytical case, because in Subsection 3.3.2 it was shown that, starting with the class of regularity C^2, the generatrices in asymptotic parametrization have directions composing a family of class $C^{0,1}$, so the left-hand side of the equation (4.20) has a meaning in a general sense.

Using the system (4.20), we can obtain a new proof of Proposition 4.1 already without assuming the analyticity of the sought-for surface (but assuming the absence of the tangency of the generatrix and the midline). Indeed, let us assume that the directrix is plane with $z = const$. Then the third equation of the system (4.20) becomes homogeneous, and it cannot have a nontrivial solution with its vanishing in at least one point, which is necessary, as $c(\varphi)$ is 2π-antiperiodic. This means that $c(\varphi) \equiv 0$, and in this case the normal to the surface has the direction $\{0, 0, 1\}$, i.e., the surface is plane and orientable.

Let us return to finding a general form for the solution of the system (4.20). We have:

$$\mathbf{l}(\varphi) = A(\varphi)(\mathbf{C} + \mathbf{B}(\varphi)), \tag{4.21}$$

where

$$A(\varphi) = exp(\int_0^{\varphi} \mu(\theta)d\theta), \quad \mathbf{B}(\varphi) = \int_0^{\varphi} \frac{\nu(\theta)\mathbf{r}'(\theta)}{A(\theta)}d\theta),$$

and $\mathbf{C}$ is some constant three-dimensional vector. Choose this vector $\mathbf{C}$ such that the vector of the direction of the generatrices $\mathbf{l}(\varphi)$ be

2π-antiperiodic. We have:

$$A(\varphi + 2\pi) = A(2\pi)A(\varphi), \ \mathbf{B}(\varphi + 2\pi) = \mathbf{B}(2\pi) - A^{-1}(2\pi)\mathbf{B}(\varphi),$$

therefore,

$$\mathbf{l}(\varphi+2\pi) = A(2\pi)A(\varphi)\left(\mathbf{C} + \mathbf{B}(2\pi) - \frac{\mathbf{B}(\varphi)}{A(2\pi)}\right) = -A(\varphi)(\mathbf{C}+\mathbf{B}(\varphi)).$$

From this equality we have that

$$\mathbf{C} = -\frac{A(2\pi)}{1 + A(2\pi)} \int_0^{2\pi} \frac{\nu(\theta)\mathbf{r}'(\theta)}{A(\theta)}\,d\theta.$$

From here, we have the final form of the solution

$$\mathbf{l}(\varphi)=\frac{A(\varphi)}{1+A(2\pi)}\left(\int_0^\varphi \frac{\nu(\theta)\mathbf{r}'(\theta)}{A(\theta)}\,d\theta - A(2\pi)\int_\varphi^{2\pi}\frac{\nu(\theta)\mathbf{r}'(\theta)}{A(\theta)}\,d\theta\right).$$

$$(4.22)$$

We complete the performed study by the following:

Theorem 4.7. *If the directrix $\mathbf{r}(\varphi)$ of the sought-for Möbius band with the generatrices non-tangent to the directrix is known, then its generatrices of necessity have the form yielded by the formula (4.22), where $\mu(s)$ and $\nu(s)$ are, respectively, the antiperiodic and periodic functions, which should be chosen such that the directions $\mathbf{l}$ be nowhere tangent to the directrix.*

To obtain a concrete flat Möbius band, consider a particular case, when in (4.20) the coefficient $\mu = 0$. Choose the directing curve Γ such that its projection to the plane xy be a unit circumference, and that it itself lie on the surface $z = xy$. This curve has the position vector $\mathbf{r}(\varphi) = \{\cos\varphi,\ \sin\varphi,\ z = \sin\varphi\cos\varphi\}$. Further, we choose ν as $\sin(\varphi/2)$ – the simplest function with the required property of 2π-antiperiodicity. Using the formula (4.22), we obtain

$$\mathbf{l} = \{\frac{1}{3}\sin\frac{3}{2}\varphi - \sin\frac{\varphi}{2},\ \cos\frac{\varphi}{2} - \frac{1}{3}\cos\frac{3}{2}\varphi,\ \frac{1}{3}\cos\frac{3}{2}\varphi - \frac{1}{5}\cos\frac{5}{2}\varphi\}.$$

The direct verification shows that $[\mathbf{l}, \mathbf{r}'] \neq 0$, i.e., the generatrices are nowhere tangent to the curve Γ, which is what is required by

I.Kh. SABITOV

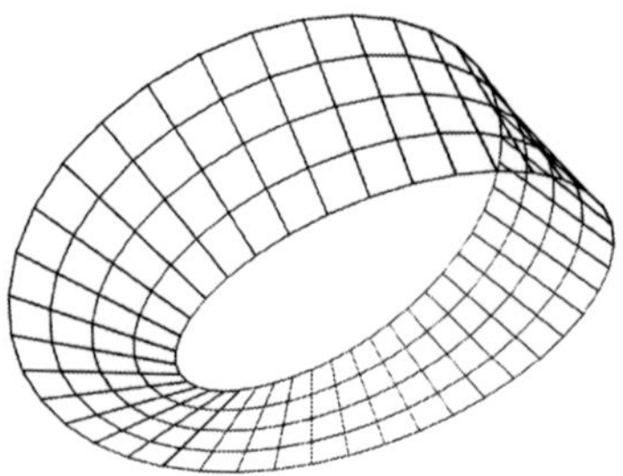

Figure 14 (courtesy of Izvestiya: Mathematics)

the condition of constructing the band. An idea of the shape of the surface S : $\mathbf{R}(\varphi, t) = \mathbf{r}(\varphi) + t\mathbf{l}(\varphi)$ can be obtained from Figure 14. This Möbius band lies on an algebraic surface of 7th order with no more than the three-valued projection to the plane (x, y). For the explicit form of its equation, see [146].

In a sense, this surface is the simplest, as its midline is the intersection of two surfaces of the second order $z = xy$ and $x^2 + y^2 = 1$ (a lower order for at least one surface is out of the question, as the midline could not be plane), and the directions of the generatrices are chosen using trigonometric polynomials of the least admissible degrees. But this surface S is not an isometric embedding of a rectangular Möbius strip into E^3. To be able to visualize the plane domain D, to which the surface S is isometric, let us find the pre-image of its midline Γ. For this, let us make use of the fact that the geodesic curvature of the midline is the ordinary curvature of its pre-image γ on the plane. Calculating the geodesic curvature $k_g(\Gamma)$ and then finding by the curvature $k_\gamma = k_g(\Gamma)$ the equation of the curve γ on the plane (ξ, η), by the known formulae we obtain the parametric representation $\xi(\varphi), \eta(\varphi)$ of the curve γ in quadratures (which, unfortunately, do not admit calculations in elementary functions). Numerical integration yields for γ the form of the curve shown in Figure 15, whereby the domain isometric to S is some band along γ, see Figure 16 (it is important to note that the strip was cut such that at the endpoints of the curve γ the generatrix AB be orthogonal to the curve and to the edge of the strip, otherwise we would obtain

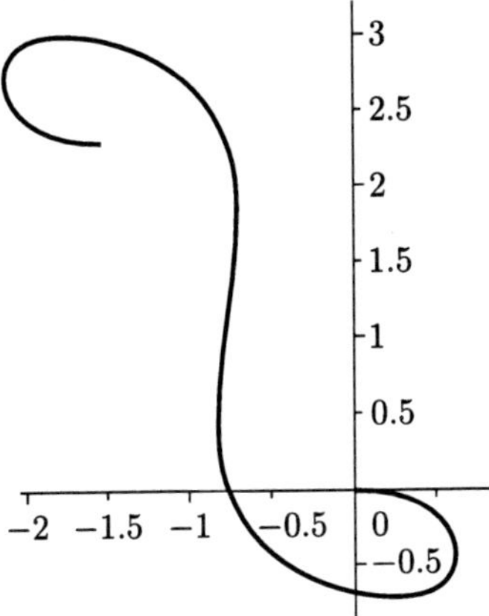

Figure 15 (courtesy of *Izvestiya: Mathematics*)

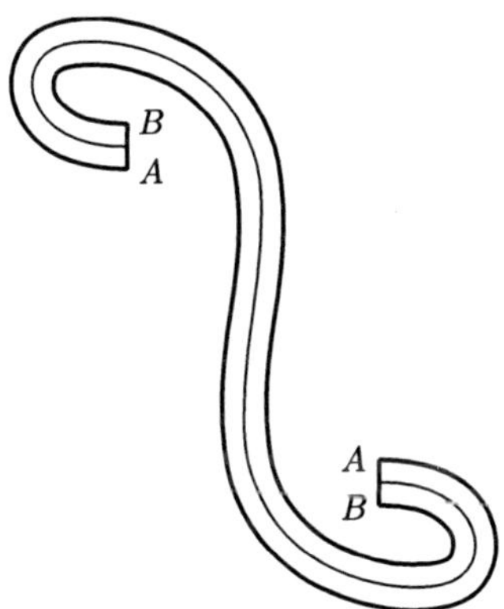

Figure 16 (courtesy of *Izvestiya: Mathematics*)

no smooth surface in identification). These computations can be also performed using the method of reducing the metric of a developable surface to the standard form presented in Subsection 2.2.3 (d).

The performed calculations give grounds to put the following question:

Open question 4.3. How, by the known algebraic equation $F(x, y, z) = 0$ of a developable surface with the known algebraic equation of some its directrix Γ, to find the algebraic or, in a general case, analytical equation $f(\xi, \eta) = 0$ of the pre-image γ of this directrix in a respective domain on the Euclidean plane (ξ, η)?•

Open question 4.4. A more general problem is the description of all algebraic surfaces with zero Gaussian curvature (certainly, not reduced to cones and cylinders) and, in particular, containing Möbius bands.•

Remark 4.19. The following remark can prove useful in solving the above posed questions: a sufficiently narrow strip $2\varepsilon > 0$ in width along the plane curve $\gamma : x = x(s), y = y(s)$ with the natural parameter s has an l. E. metric $(1 + k(s)t)d^2s^2 + dt^2$, where $k(s)$ is the curvature of the curve γ, and the parameter t, $|t| \le \varepsilon$, defines for each point γ the lengths laid off within the limits of the strip along the straight lines orthogonal to γ, so that the l. E. metric of the sought-for Möbius band, which realizes the strip along the curve γ as a pre-image of the midline Γ of the Möbius band, can be considered to be defined in the rectangle $0 \le s \le L, |t| \le \varepsilon$, where L is the length of the curve γ.•

4.3.3.5 Möbius bands with the generatrices orthogonal to the midline. Let us continue to study the possible types of the flat Möbius band with generatrices not tangent to the midline. As in this case the product $(\mathbf{lr}')$ is antiperiodic, it vanishes in at least one point. This means that on any flat Möbius band there exists a generatrix orthogonal to the midline of the surface. On the known non-orientable surface

$$x = \cos\varphi + t\cos(\varphi/2)\cos\varphi, \quad y = \sin\varphi + t\cos(\varphi/2)\sin\varphi,$$
$$z = t\sin(\varphi/2), \quad (0 \le \varphi \le 2\pi, -1/3 \le t \le 1/3). \tag{4.23}$$

with negative curvature, found by Maschke [107], all generatrices are orthogonal to the midline. It proves that for the standard flat Möbius band the situation is opposite. This is confirmed by the following property of flat Möbius bands.

Proposition 4.2. *There exists no C^2-smooth standard Möbius band, on which the generatrices are everywhere orthogonal to its midline.*

The proof is very simple. Indeed, if such a band $S : \mathbf{r}(s)+t\mathbf{l}(s), |\mathbf{l}| = 1$, exists, the coordinate lines $t = const$ and $s = const$ are geodesics, the lengths of the arcs on which are equal to the lengths of respective segments forming on the pre-image (i.e., on the rectangular Möbius strip) some rectangle. Herewith, the pre-images of the generatrices will be parallel segments of the straight lines orthogonal to the midline of the rectangle, therefore their images – the generatrices on the surface – will also be parallel, which means that S is a cylindrical surface. Then the directrix $t = 0$ due to the uniqueness of the geodesic in this direction will coincide with the plane line obtained in the section of the surface by the plane orthogonal to the generatrices, and this, as we know, under the condition of nontangency of the generatrices and the directrix, is impossible even in class C^2.

Let us see if any Möbius band with generatrices orthogonal to the midline will exist.

We will perform all reasoning assuming that the midline Γ is given by its natural equation, so the parameter of the curve – the length of its arc – will be designated as s, $0 \leq s \leq L$, where L is the length of the midline. We know that on a regular analytical curve we can introduce analytical fields of the main normal $\mathbf{N}(s)$ and binormal $\mathbf{B}(s)$ and the analytical functions of curvature $k(s) = k_A(s)$ and torsion $\varkappa(s)$. The cases of vanishing of the function of curvature are not excluded, but the normal to the surface does not have to be parallel to the principal normal, because now the midline is not a geodesic. Let all generatrices be orthogonal to Γ. We can choose their directrices as unit vectors $\mathbf{l}(s)$. By multiplying the equation (4.20) scalarwise by $\mathbf{l}$, we obtain $\mu = 0$. This means that now $\mathbf{l}(s)$ and $\mathbf{r}(s)$ are related by the equation

$$\mathbf{l}'(s) = \nu(s)\mathbf{r}'(s). \tag{4.24}$$

As $(\mathbf{l}\mathbf{r}') = 0$, we have the representation

$$\mathbf{l} = \cos \alpha \mathbf{N} + \sin \alpha \mathbf{B}, \tag{4.25}$$

where $\mathbf{N}$ and $\mathbf{B}$ are, respectively, the principal normal and the binormal of the curve Γ. We substitute the derivative of $\mathbf{l}$ into the equation

(4.24). With account for the Frenet equations, we obtain

$$\mathbf{l}' = -\sin\alpha\,\mathbf{N}\alpha' + \cos\alpha(\varkappa\mathbf{B} - k_A\mathbf{r}') + \cos\alpha\,\mathbf{B}\alpha' + \sin\alpha(-\varkappa\mathbf{N}) = \nu\mathbf{r}',$$

wherefrom we have $\nu = -k_A\cos\alpha$, $\alpha' = -\varkappa$. Therefore,

$$\alpha(s) = -\int_0^s \varkappa(\sigma)d\sigma + C. \tag{4.26}$$

The vector $\mathbf{l}(s)$ should be antiperiodic. Depending on the periodicity or antiperiodicity of the vectors $\mathbf{N}$ and $\mathbf{B}$ we may have two cases:

1) $\mathbf{N}$ and $\mathbf{B}$ are periodic, then for the antiperiodicity of $\mathbf{l}(s)$ it is necessary and sufficient that the condition $\alpha(s+L) = \alpha(s)+\pi+2\pi p$, where p is an integer, be satisfied. This condition leads to an equality of the form

$$\int_0^L \varkappa(s)ds = \pm\pi + 2\pi p, \tag{4.27}$$

upon satisfaction of which the relations (4.26) and (4.25) define the vector field of the directions $\mathbf{l}(s)$, setting the flat Möbius band with the generatrices orthogonal to the midline.

2) $\mathbf{N}$ and $\mathbf{B}$ are antiperiodic, then for the antiperidicity $\mathbf{l}(s)$ it is necessary and sufficient that the condition $\alpha(s + L) = \alpha(s) + 2\pi p$, where p is some integer, be satisfied. This condition can be written as the requirement of satisfying the equality of the form

$$\int_0^L \varkappa(s)ds = 2\pi p, \tag{4.28}$$

at which the relations (4.26) and (4.25) define the vector field of directions $\mathbf{l}(s)$, setting the flat Möbius band with the generatrices orthogonal to the midline.

In both cases the choice of these generatrices is not unique but is defined by the freedom in the choice of the constant C in the formula (4.26). In particular, if, choosing $C = 0$, we have the surface S_0, then at $C = \pi/2$ the respective generatrices will be the normals to S_0, therefore, on S_0 the curve Γ will be the line of curvature, and it will be the line of curvature on all surfaces S_C, only with different normal curvature. Therewith, we obtain that on the basis of one directrix Γ we can construct a continuous family of Möbius bands with the parameter C, on which the generatrices are orthogonal to the directrix.

The real existence of such Möbius bands is shown thus. For the second case, it is the simplest to use the line (4.10) taken from a work by Schwarz ([156]), because it is already known that on it the principal normal and the binormal are antiperiodic. Calculating its integral torsion, we obtain that it is approximately equal to (-18). Let us consider the family of curves $\Gamma_\lambda : x = \sin\varphi, y = \lambda(1 - \cos\varphi)^3, z = (1 - \cos\varphi)\sin\varphi$. These curves are without self-intersections; their curvatures and torsions have the same zeros of the same orders as the initial Schwarz curve, and in continuous homotopy their principal normals and binormals remain antiperiodic. At $\lambda \to 0$ their torsion tends to zero, being positive, so there will be such values of the parameter λ, at which the integral torsion of respective curves will be equal to -4π and -2π. Therefore, these curves can serve as the midline for the Möbius band with the generatrices orthogonal to this line.

To find the sought-for curves in the first case, we will make use of the idea (given in [86], p. 146) of constructing curves with the predetermined value of integral torsion. Consider two cylindrical helices

$$L_1 : \ x = \cos\varphi, y = \sin\varphi, z = h\varphi, 0 \le \varphi \le (2n + 1)\pi,$$

and

$$L_2 : \ x = \cos\theta, y = \sin\theta, z = H(2\pi - \theta), \pi \le \theta \le 2\pi.$$

Let us choose a natural number n and positive parameters h and H such that it will be $H = (2n + 1)h$ and

$$\frac{(2n + 1)h\pi}{\sqrt{1 + h^2}} - \frac{H\pi}{\sqrt{1 + H^2}} > (2m + 1)\pi, \qquad (4.29)$$

where $m < n$ is some fixed non-negative integer (these conditions mean that the curves L_1 and L_2 occur at the points $M_1(1, 0, 0)$ and $M_2(-1, 0, H\pi)$ and their total integral torsion is greater than $(2m + 1)\pi$). By elementary calculations, it is verified that the inequality (4.29) is satisfied at fixed $h > 0$ and m and a sufficiently large n.

The curve $L = L_1 + L_2$ is closed, nonsmooth at two points and self-intersecting, but by an arbitrarily small change of its segments around the points M_1 and M_2 the curve can be smoothed to a regular curve of class C^∞, and the self-intersection can be eliminated by transferring the main part L_2 to a cylinder of some radius $1 + \varepsilon$ with arbitrarily small $\varepsilon > 0$. Under these transformations, integral torsion will change little (as we do not need to take into account the derivative of torsion, and at small h and sufficiently large n the torsion of the curves differs little) and will remain greater than $(2m + 1)\pi$. Let us denote the thus obtained curve of smoothness C^∞ as $\tilde{L}$. We substitute now the periodic components of the position vector of this curve by partial sums of their Fourier series such that integral torsion of the obtained analytical curve L_* remain greater than $(2m + 1)\pi$ (this is possible because the Fourier series converge with derivatives of any order). Let

$$x_*(s),\, y_*(s),\, z_*(s)$$

be components of the position vector of the analytical curve L_*. Consider, as above, the family of the curves L_λ with the equation $x = x_*(s), y = y_*(s), z = \lambda z_*(s)$. Again from the formulae for torsion we will see that integral torsion of the curves L_λ at $\lambda \to 0$ tends to zero, which means that the values of the parameter, at which integral torsion is equal to the defined value of the form $(2k + 1)\pi$ with any natural $k \leq m$, which is what proves the existence of the sought-for Möbius bands also in the case of periodic normals to the midline.

4.3.3.6 Some unsolved problems. The work [193] puts the following:

Open question 4.5. What is the relation between the length l and height h of the standard rectangular Möbius strip, which admits an isometric embedding into E^3?•

The work [70] proves that there exists such $\varepsilon > 0$ that an embedding is possible only at $l > (\frac{\pi}{2} + \varepsilon)h$, and *hypothesizes* that *at* $l \leq \sqrt{3}h$ *there is no embedding.* Therewith, it is expected that $\varepsilon = \sqrt{3} - \pi/2$; this expectation is supported by the fact that if smooth embeddings containing flat pieces are considered, then at $l > \sqrt{3}h$

such embeddings do exist[35]. But if we search not for smooth but piecewise linear embeddings, they exist at any ratio of the length and height of a rectangle, see [13, 33] (but the Gardner realization of an arbitrary Möbius band mentioned in [157] and described in the previous footnote is not an embedding or even an immersion).•

Open question 4.6. Can we assert that on any Möbius band with a flat metric there exists a closed directrix playing the role of the midline and nowhere tangent to the generatrix?•

Open question 4.7. There are many more simple plane domains, whose isometric realization in R^3 yields geometrically interesting surfaces. For instance, we can put a question on the realization of a domain in the form of a cross (similar to the symbol of the International Red Cross) with its identification in pairs of opposite sides in three variants (the identification in both pairs is done as during the glueing of a cylinder, or as during the glueing of a Möbius band, or one pair is glued as in a cylinder and the other as in a Möbius band).•

Remark 4.20. By the way, the concept of the midline of a Möbius band remains rather indefinite, as the general case assumes only the diffeomorphism of the Möbius band to the standard flat Möbius band, and the identified endpoint lines of the natural development of an arbitrary Möbius band (analogues of the lines AB and BA from Figure 16) can be arbitrary arcs, at that, not always orthogonal at the ends of the band (but, certainly, with such a sum of corresponding angles at two ends, which provides for the smoothness of the glued band). At the very beginning of Subsection 4.3.3, we gave a definition of the

---

[35] Imagine a rectangular Möbius band with $l = \sqrt{3}h$ in the form of an isosceles trapezoid $ABCD$, where the longer base $|AD| = 2|BC|$, the angle $\angle BAD = 60°$, and the sides AB and CD are subject to identification with $A = C, B = D$. Then break the trapezoid into three regular triangles ABK, BKC, CKD, where K is the middle of the side AD. By bending the trapezial band of paper along the segment KB to $180°$, bring the vertex C into coincidence with the vertex A and then, by turning the triangle DKC around the segment CK, bring the vertices D and B into coincidence. Now it is clear that if it were $l > \sqrt{3}h$, the bends along the segments BK and KC could be replaced with narrow cylindrical strips, smoothly changing into flat triangles, and an embedding into E^3 could be obtained.

midline as an image in the isometry or diffeomorphism of the mid-line of a rectangular Möbius band, but it remains unclear whether it is the most satisfactory in the case of ordinary Möbius bands. The midline can also be defined as a smooth closed curve, during the walk along which the normal to the band changes its direction. Such lines would, certainly, be many, so the midline of the shortest length can be called the *main midline.*●

An immediate question arises:

Open question 4.8. Does the main midline exist on any Möbius band and is it unique?●

4.3.4   *Isometric immersions of two-dimensional flat metrics into a hyperbolic space H^3*

The extrinsic curvature of a surface with a flat metric in H^3 is equal to $+1$, so such surfaces are locally convex there. This means that a torus and non-orientable manifolds with an l. E. metric cannot be immersed into H^3. As for a plane and a cylinder, they admit an isometric embedding into H^3 – respectively as an orisphere and by the equidistant surface of a geodesic, see [155, 166, 186]. The proof is based on the Jörgens theorem [91] on the solutions on the plane of the equation

$$z_{xx}z_{yy} - z_{xy}^2 = 1, \qquad (4.30)$$

which states that such solutions are polynomials of degree 2. For surfaces in H^3 with a flat metric, [59] gives a representation analogous to the Weierstrass formulae for minimal surfaces, and, based on this, obtains a new proof of the above Volkov–Vladimirova–Sasaki theorem and investigates new classes of incomplete surfaces of zero curvature in H^3, in particular, surfaces of revolution; and gives descriptions of the ends of surfaces corresponding to the solutions of the equation (4.30) in domains with a punctured point. In [62] the equation (4.30) is considered on the plane with n punctured points; the work gives a complete description of the solutions, for which these points are unremovable features (the case of $n = 1$ was considered earlier in [92]). Based on this description, we can obtain new classes of surfaces in H^3 with an l. E. metric. We can also mention the work [60] dealing with immersions of l. E. metrics into pseudo-Euclidean space L^4 in the form of surfaces with flat normal connection. The general theory of such immersions constructed in it also enables the description of surfaces with flat metrics and in three-dimensional hyperbolic space considered as a respective hypersurface in L^4.

5   Isometric immersions of l. E. metrics into E^4

5.1   Local theory of isometric immersions of two-dimensional flat metrics into E^4

The general theorem on the isometric immersions of two-dimensional metrics into E^4, belonging to E.G. Poznyak [127], affirms that if an arbitrary-sign complete $C^{3,\alpha}$-smooth metric is given on the plane, then any simply connected compact domain of the plane admits an isometric immersion into R^4 by some surface of class $C^{2,\alpha}, 0 < \alpha < 1$. In [142], it has been shown that this theory can be sharpened as follows: first, the condition of completeness of the metric should be removed; second, in respect of the smoothness, it can be affirmed that the $C^{n,\alpha}$-smooth ($n \geq 2, 0 < \alpha < 1$) metric is immersed into R^4 by the surface of the same smoothness $C^{n,\alpha}$ and, according to [150], this result in smoothness cannot be improved.

As for immersions of l. E. metrics into E^4 and surfaces with l. E. metrics into E^4, there are, certainly, their own specific features here. The local structure of isometric immersions of l. E. metrics into E^3 has been studied fully enough, which, however, cannot be said of their immersions into E^4. Surfaces in E^4 with l. E. metrics do not, have such a simple structure as in E^3, so some additional requirements should be *a priori* imposed on their external structure. Below, in Subsection 6.3.3, we will describe even at general values of dimensions and codimensions the cases of ruled structure of surfaces with l. E. metrics; now we will only briefly give an account of what is known of these surfaces in E^4. The main results in this direction (including the problems of the structure of surfaces in the large) were obtained in [17, 18, 45, 47, 50–53, 55, 56, 61, 76, 97, 99, 116, 122, 127, 154, 166, 167, 170, 176, 187, 189, 191, 194, 195] and other works (see also the reviews in [24], §5.2 of [26], [69]; note that [53] points to one error in [191]). It is first of all clear that any developable surface from R^3 can be placed into R^4, isometrically immersing for this all the space of R^3 (or its part containing inside itself the surface considered in E^3) into E^4, e.g., by "folding" E^3 into the three-dimensional cylinder located in E^4 (for this, it is sufficient to take an isometric immersion $h : (X, Y, Z) \rightarrow (\cos X, \sin X, Y, Z) \in E^4$, which will transfer the surface $(X = x(s,t), Y = y(s,t), Z = z(s,t)) \subset E^3$ into a surface isometric to it in E^4). Such immersions are said to be obtained by composition. The exact definition of a composition is given as follows: an immersion $f : (U \subset \mathbb{R}^2) \rightarrow \mathbb{R}^4$ is called a *composition*, if there exist isometric immersions $g : (U \subset \mathbb{R}^2) \rightarrow V \subset \mathbb{R}^3$ and $h : (V \subset \mathbb{R}^3) \rightarrow \mathbb{R}^4$, such that $f = h \circ g$ (herewith, a mapping h is called an *isometric extension* of the mapping g, and the mapping g is called *isometrically extended*)[36]. Certainly, in the first instance it is of interest to exclude compositions from consideration as, in a sense, they are trivial immersions (such immersions were singled out into a special class earlier in [194] and [195], where they are assigned the concept of a surface obtained by *co-bending* – a surface in some space V passes into an isometric surface in another space W of greater dimension, co-bending, i.e., bending together, with ambient space V it contains, in the process of its isometric mapping into W). It turns out that the singling out of isometric immersions $f : (U \subset R^2) \rightarrow R^4$, being compositions, and in general the description of surfaces with an l. E. metric, are closely related to the behaviour of their first normal space N_1^f consisting of the linear span of the image of the vector-valued second quadratic form of the surface. Namely, the constancy of the dimension of this space is important (it is not without reason that surfaces of such a type are called in [166] "nicely curved surfaces"). In a three-dimensional variant, this corresponds to the condition that a developable surface in E^3 is either plane everywhere, or has no plane domains and even generatrices with points of flattening, "interfering" with the uniformity of type of the structure of the surface (as cylindrical, conical or torsial ones). Let M_2 be some two-dimensional Riemannian manifold. Assume that the first normal space of an isometric immersion $f : M^2 \rightarrow E^4$ has dimension 1 everywhere and is nonparallel in normal connection of the surface. For

---

[36] In [48], we can find some description of surfaces obtained by composition in R^n due to embeddings $R^m, m < n$, into R^n.

this case, [51] proves that then the manifold M^2 of necessity has a flat metric, and the immersion f is obtained as the composition $f = F \circ i$, where i is a completely geodesic immersion of M^2 into E^3 (i.e., its image – simply a domain on some plane in E^3), and F is an isometric immersion of E^3 into E^4 without points of flattening[37]. This assertion, as its authors write, gives an unexpectedly simple description of the geometry of such surfaces, but we want to illustrate this by an example to show that in reality the idea of the obligatory existence of the decomposition $f = F \circ i$ is not at all evident. Consider the immersion of the Euclidean metric $ds^2 = du^2 + dv^2$ into R^4 given by the formula $f : x_1 = \cos(\cos u), x_2 = \sin(\cos u), x_3 = \sin u, x_4 = v$ in some neighbourhood D of the point $(u_0, v_0) = (0,0)$. This immersion was obtained as a composition of immersions $g : D \to R^3$ and $F_1 : R^3 \to R^4$, where $g : (u, v) \to (x, y, z) = (\cos u, \sin u, v)$, and $F_1 : (x, y, z) \to (x_1, x_2, x_3, x_4) = (\cos x, \sin x, y, z)$. Both immersions g and F_1 specify in their spaces cylinders without points of flattening, but in the composition $f = F_1 \circ g$ the image of the mapping g does not lie on a completely geodesic submanifold in R^3. Meanwhile, all conditions of the respective theorem from [51] are satisfied. Indeed, in the normal space of the surface $f(D) \subset R^4$ we can choose the following unit normals:

$$\mathbf{n}_1 = \frac{\tilde{\mathbf{n}}_2 - \tilde{\mathbf{n}}_1 \sin^2 u}{\sqrt{1 + \sin^4 u}}, \quad \mathbf{n}_2 = \frac{\tilde{\mathbf{n}}_1 + \tilde{\mathbf{n}}_2 \sin^2 u}{\sqrt{1 + \sin^4 u}},$$

where

$$\tilde{\mathbf{n}}_1 = \{(\cos(\cos u), \sin(\cos u), 0, 0)\},$$
$$\tilde{\mathbf{n}}_2 = \{(\sin(\cos u) \cos u, - \cos(\cos u) \cos u, - \sin u, 0)\}.$$

These normals are assigned the following second forms of the surface:

$$II_1 = \sqrt{1 + \sin^2 u}\, du^2, \quad II_2 = 0,$$

---

[37] In reality, [51] obtained a stronger result on the isometric immersions $f : M^n \to Q_c^m, 2 \le n < m - 1$, where Q_c^m is a space of constant curvature c and dimension m, asserting that if the dimension of the first normal space of the surface $f(M^n)$ is equal to 1, then the immersed manifold M^n also should have a constant curvature c, and the immersion f itself proves to be a composition of the same kind $F \circ i$, as in the case of the immersion of M^2 into E^4.

so that the dimension of the first normal space N_1^f is 1; besides, the projection of the covariant (in R^4) derivative of the normal n_1 to the normal space does not belong to N_1^f, therefore N_1^f is not parallel in normal connection.

The representation of the mapping f in the required form is achieved as follows[38]. From Subsection 2.2.3 we know that any curve in R^3 (irrespective of whether it is plane or spatial), having a non-zero curvature, can be a geodesic on some developable surface S. Let us consider in R^3 the curve $\Gamma : \rho = \rho(u) = \{\cos(\cos u), \sin(\cos u), \sin u\}$ with the natural parameter u and construct for it a corresponding developable surface S. In Subsection 2.2.3, it is shown that the position vector $\mathbf{r}$ of this surface can be written in the function of (x, y) such that it was $ds^2(x, y) = dx^2 + dy^2$ and that at $y = 0$ it was $u = x$, such that $\mathbf{r}(x, 0) = \rho(u)$. Now the mapping $f : D \to R^4$ can be written down as $f(u, v) = F(x, y, z) \circ i$, where $i : D \ni (u, v) \to (u, 0, v) \in R^3$, $F : R^3 \ni (x, y, z) \to (r(x, y), z) \in R^4$. The performed construction gives an idea of the proof of the general proposition.

But if the first normal space, having dimension 1, is parallel in normal connection of the surface, then $f(M_2)$ is contained in reality in some three-dimensional plane of a four-dimensional space[39], so in the case of $M_2 = U \subset R^2$ the surface $f(M_2)$ proves to be a usual developable surface in $E^3 \subset E^4$.

Let it now be known that the first normal space of an isometric immersion $f : (U \subset E^2) \to E^4$ has dimension 2. Such surfaces, in turn, differ in whether the normal connection on them is plane or not, i.e., whether the curvature $R^\perp$ of normal connection is equal to

---

[38] By my request, this part of the example was completed by the Spanish mathematician J. Gálvez, and I use this occasion to extend my sincere thanks and appreciation to him.

[39] This proposition also takes place in the case of general dimensions and co-dimensions in the following formulation: for the surface $f : M^n \to M_c^m$ with $dim N_1^f = r$ to prove located in some totally geodesic submanifold $Q_c^{n+r} \subset Q_c^m$, it is necessary and sufficient that its first normal space be parallel in normal connection. In other terms, this proposition and various (associated with parallel normal sub-bundles) properties of the submanifolds can be found in [37, 41, 47, 106, 166] and in other works cited therein. Other features of the assignment of a two-dimensional surface from E^4 to space E^3 can be found in [8] and references therein.

zero or not. The description of surfaces with $dim(N_1^f) = 2$, having a flat metric and a flat normal connection, is dealt with in [52]. In the domain $U \subset E^2$, this work introduces four functions (h_1, h_2, v_1, v_2), where h_1 and h_2 are given via two arbitrary functions $\Phi_1(t)$ and $\Phi_2(t)$ of one argument by the following formulae:

$$h_1(u, v) = \Phi_1(u+v) + \Phi_2(u-v), \quad h_2(u, v) = -\Phi_1(u+v) + \Phi_2(u-v),$$

while $v_1(u, v)$ and $v_2(u, v)$ are defined via some linear system of two differential equations with participation of h_1 and h_2 in the coefficients of the system, the solution of which in turn depends on two functions of one argument. It is being proved that such a set of functions under the condition that $v_1 \neq 0, v_2 \neq 0$ defines some isometric immersion $f : U \to R^4$, and vice versa, any isometric immersion $f : U \to R^4$ is assigned four functions of one argument, which determine, by the above described rule, the functions h_1, h_2, v_1, v_2. In addition to this, a criterion is being established, which makes it possible to characterize among these functions those which correspond to the cases, when the isometric immersion $f : U \to R^4$ is a composition (an example of the product of two plane curves $(x(s), y(s)), (X(t), Y(t))$ of curvature not equal to zero shows that the considered class of surfaces does include compositions:

$$f : (s, t) \to (x(s), y(s), X(t), Y(t)) = F \circ g,$$
$$g(s, t) = (x(s), y(s), t), F(x, y, z) = (x, y, X(z), Y(z)).$$

Thus, it can be considered that in a sense the issue of the local description of isometric immersions of l.E. metrics into E^4 with $R^\perp = 0$ and $dim(N_1^f) = 2$ is solved, including the singling-out of the class of immersions in the form of convolutions.

The works [61] and [189] also give a description of isometric immersions of flat metrics into E^4, but already without the assumption of locality, including for the immersion of all the plane E^2 (in the same assumptions of $R^\perp = 0$ and $dim N_1^f = 2$). They widely apply the method associated with the use of Gaussian mapping of a surface into a Grassmannian manifold $G_{2,4}$. The finding of isometric immersions of E^2 and simply connected domains in it into E^4 is reduced to the solution of some linear hyperbolic system of differential equations. Their solution also has, as in [51], an arbitrary rule in the choice of four functions of one variable.

The work [53], in terms of necessary and sufficient conditions, describes locally isometric immersions of l. E. metrics into E^4 with $R^\perp \neq 0$ not being compositions; what is more, it turns out that all these immersions can be in a rather complicated way related to some isometric immersion of l. E. metrics into the sphere $S^3 \subset E^4$.

The results of these works provided the grounds to advance a hypothesis (in [26], pp. 410 and 464) that *any two-dimensional compact surface in E^4 with an l. E. metric and with flat normal connection is either a torus in $S^3 \subset R^4$ or a Cartesian product of two plane closed curves, each of which lies on a two-dimensional plane orthogonal to the other plane.* But in [61] this hypothesis was shown to be false. Below, we will discuss this in greater detail.

5.2   *Local immersions of flat metrics into S^3*

Practically in all works related to the immersion of l. E. metrics into R^4, a special place is occupied by their immersions into the three-dimensional sphere $S^3 \subset R^4$. These immersions are remarkable in that locally they admit the same complete description as developable surfaces in E^3. A common feature for them, as for immersions into E^4, are the properties of $dim(N_1^f) = 2$ and $R^\perp = 0$, which is easy to see by the derivation formulae and Weingarten equations for surfaces in elliptic space adduced, e.g., in the book [124], Chapter V, §1. The structure of these surfaces was studied long ago by Bianchi [17, 18], and henceforth, particularly beginning with [166, 167], was the subject of studies by many authors. In the sphere, the extrinsic curvature of surfaces with zero Gaussian curvature is negative (and equal to $-\dfrac{1}{R^2}$, where R is the radius of the sphere, which in the following we consider to be equal to 1)[40], so on them there are two families of asymptotic lines. In the asymptotic coordinates (u, v), the first form has the presentation $ds^2 = du^2 + 2\cos\omega\, dudv + dv^2$, and the

---

[40] In the general case, in the spherical space of radius R for the extrinsic curvature K_{ext} of the surface in standard classical notation, there is the formula
$$K_{ext} = k_1 k_2 = \frac{LN - M^2}{EG - F^2} = K_{int} - \frac{1}{R^2}.$$

second one is equal to $\pm 2 \sin \omega \, du dv$, where the angle ω between the coordinate asymptotic lines can be represented as $\omega(u, v) = U(u) + V(v)$ (as $\omega_{uv} = -K \sin \omega = 0$). In these coordinates, the domain of definition of the respective standard form $dx^2 + dy^2$ of the metric is determined by the mapping $x = x(u, v), \ y = y(u, v)$, where

$$x = \int \cos U(u) du + \int \cos V(v) dv$$

$$y = \int \sin U(u) du - \int \sin V(v) dv \qquad (5.1)$$

(see Subsection 2.2.3(b)). Besides, by the Beltrami–Enneper theorem, also valid in spherical space, the torsion of an asymptotic line on a surface with $K_{ext} = -1$ is equal to ± 1 (if the asymptotic is simultaneously a geodesic, then its torsion *by definition* is taken to be equal to ± 1; the choice of sign will be made more exact somewhat later). In the considered metric, the geodesic curvature of the line $v = const$ ($u = const$) is equal to $-U'(u)$ ($V'(v)$), therefore, it does not depend on the value of v (u). As the geodesic curvature of the asymptotic curve is equal to its curvature as a curve in space S^3, we have that all asymptotic u-lines (v-lines) have the same curvatures and torsions in the arc function, which means that they are congruent. Therefore, *any surface of zero curvature in S^3 is a translation surface of one asymptotic curve along another asymptotic.* In the spherical space, analogues of Euclidean translations are motions at which all points of a sphere move away from their initial position to the same distance. There are two kinds of such motions. In the representation of motions in the sphere considered as a surface in R^4, these motions are assigned the action on a point of the sphere by special orthogonal operators, whose matrices make some subgroup in $O^+(4)$. In the quaternion representation of the sphere S^3 as a Lie group of quaternions $q = q_1 + iq_2 + jq_3 + kq_4$ with the module $|q| = 1$ (by multiplication further denoted by the symbol $\cdot$) they are assigned the action of an element of the group from the left or from the right. Recall that the action of the element $a = a_1 + a_2 i + a_3 j + a_4 k$ from the left on the element $b = b_1 + b_2 i + b_3 j + b_4 k$ can be written in a matrix form as the product $A(a)b^T$, and the action of the element b from the right on a as the product $aB(b)$, where the matrices $A(a)$ and $B(b)$ are given as follows:

$$A(a) = \begin{pmatrix} a_1 & -a_2 & -a_3 & -a_4 \\ a_2 & a_1 & -a_4 & a_3 \\ a_3 & a_4 & a_1 & -a_2 \\ a_4 & -a_3 & a_2 & a_1 \end{pmatrix},$$

$$B(b) = \begin{pmatrix} b_1 & b_2 & b_3 & b_4 \\ -b_2 & b_1 & -b_4 & b_3 \\ -b_3 & b_4 & b_1 & -b_2 \\ -b_4 & -b_3 & b_2 & b_1 \end{pmatrix}$$

and their determinants are equal to $+1$. The independence of interpreting the quaternion product $a \cdot b$ as the result of the action of the left translation a to the element b or the right translation b to the element a is understood as the true equality $A(a)b^T = (aB(b))^T$ or $bA^T(a) = aB(b)$.

Let the position vector[41] $r(u, v)$ of the surface S be given in asymptotic coordinates (u, v), and let two coordinate lines

$$\gamma_1: \quad y(u) = (y_1(u), y_2(u), y_3(u), y_4(u)) \quad \text{with } v = v_0$$
$$\gamma_2: \quad z(v) = (z_1(v), z_2(v), z_3(v), z_4(v)) \quad \text{with } u = u_0$$

on it intersect in $M_0 = r(u_0, v_0) \in S$, i.e., $y(u_0) = z(v_0) = r(u_0, v_0)$; then $r(u, v)$ as the position vector of the translation surface can be represented in two ways:

$$r(u, v) = (y(u) \cdot y(u_0)^{-1}) \cdot z(v) \text{ and } r(u, v) = y(u) \cdot (z(v_0)^{-1} \cdot z(v)), \tag{5.2}$$

or, in a matrix form,

$$r(u, v) = A(u, u_0)z(v) \text{ and } r(u, v) = y(u)B(v, v_0)$$

in accordance with the above given matrix writing rules. Note that the curves γ_1 and γ_2 have a common osculating plane, which is the tangent plane of the surface; hence, in particular, it follows that at the point $M_0 = \gamma_1 \cap \gamma_2$ the binormals to γ_1 and γ_2 are collinear. The distributing of the signs of torsion between γ_1 and γ_2 will be made precise somewhat later.

---

[41] Further on, we will not mark vectors by bold letters due to their frequent identification with points of the sphere.

Generally speaking, in some cases a surface with an l. E. metric can also be constructed without specifically emphasizing its asymptotic lines. For this, let us consider the general form of the equation of translation surfaces $r(u, v) = y(u) \cdot z(v)$, where $y(u)$ and $z(v)$ are some C^1-smooth curves in S^3 referred each to its natural parameter and not tangent to each other at the pont of intersection at some values $u = u_0$ and $v = v_0$. It is easy to show that if at least one of the curves is a geodesic, i.e., a great circumference of the sphere, then the surface has an l. E. metric. Indeed, let, e.g., the curve $z(v)$ be a great circumference. By a suitable choice of the coordinate axes we can achieve its equation to take the form $(\cos v, \sin v, 0, 0)$ with $v_0 = 0$. Then for the metric of the surface $r(u, v) = y(u) \cdot z(v)$ by simple calculations we obtain the expression $ds^2 = du^2 + 2f(u)dudv + dv^2$, where $f(u) = y_1 y_2' - y_2 y_1' - y_3 y_4' + y_4 y_3'$, and the condition of nontangency in the common point $(1, 0, 0, 0)$ guarantees that $|f(u_0)| < 1$, therefore, at some neighbourhood of this point the surface $y(u) \cdot z(v)$ is regular and has the zero curvature (at a sufficient distance from the point (u_0, v_0) singular points can emerge on the surface). The translation surfaces, at which one of the transferable curves is a spherical straight line, are called *Clifford ruled surfaces*, and the straight lines themselves are called its generatrices (thereby these surfaces are in S^3 the closest analogues of developable surfaces from R^3); what is more, the inverse is also true: if some ruled surface has an l. E. metric, then it is a ruled Clifford surface). In [18], it was proved that orthogonal trajectories to the generatrices on a ruled Clifford surface are curves with torsion ± 1. Hence, it follows that if a curve with torsion ± 1 is chosen on a Clifford ruled surface as a directrix, and its generatrix is not orthogonal to it at some point, then one more curve with torsion modulo 1, orthogonal to the generatrices, will be found on the surface. And as the generatrix itself is an asymptotic line, one of these curves with unit torsion will not be asymptotic. What is more, the surface can be formed by two families of circumferences, then both curves orthogonal to the generatrices will have a torsion ± 1 and neither of them will be an asymptotic line. Such curves can be constructed, e.g., as cylindrical helices on ruled Clifford surfaces, which sometimes may prove even closed. Therefore, it can be that the surface is given in the form of $r(u, v) = y(u) \cdot z(v)$, where $y(u)$ is a curve with torsion $+1$, and the second curve is a circumference, so that the surface has a flat metric; nevertheless, it is not altogether obligatory that the curve $y(u)$ be asymptotic.

Open question 5.1. We do not know what maximum number of curves with torsion ± 1 can emanate from one point of a surface with a flat metric. This question can be put in a general form: what maximum number of curves with torsion $\varkappa = \pm\sqrt{-K}$ can pass through a point of a surface (in E^3 or S^3) with extrinsic curvature $K < 0$? In yet a more general statement, we can put a question on the existence of a curve with a known value of torsion $\varkappa(s)$ on a given surface, thereby substituting in the general theorem of the existence of a curve the condition of knowing its curvature $k(s)$ by the condition of the location of the curve on a known surface. For instance, this problem has been solved in a sense for the sphere in E^3: knowledge of the torsion of a curve on it enables finding the curvature of the curve as a two-parametric solution of the known differential equation (see, e.g., Problem 18.127 in [110]) and, thereby, to define all curves with this torsion on the sphere S^2. It appears that for surfaces in S^3 with an l. E. metric this problem is easier to investigate than even in R^3, because for them there is a standard representation of the form $y(u) \cdot z(v)$ with the known properties of "base" asymptotic curves.$\bullet$

Consider these examples. The Clifford torus

$$(x_1, x_2, x_3, x_4) = a(\cos\theta, \sin\theta, \cos\varphi, \sin\varphi),$$

written down as the *Cartesian* product of two circumferences of radius $a = \sqrt{2}/2$, situated in orthogonal planes, in asymptotic coordinates is represented in the form of a *group* product $y(u) \cdot z(v)$ of two large circumferences of the sphere S^3:

$$a(\cos u, \sin u, \cos u, \sin u) \cdot (\cos v, \sin v, 0, 0) =$$
$$a(\cos(u+v), \sin(u+v), \cos(u-v), \sin(u-v))$$

with the respective domain of the change of parameters (u, v); in this case, the angle $\omega = \pi/2$, therefore, these circumferences play the role of orthogonal trajectories of each other, so it would be impossible to obtain the existence of curves with torsion ± 1 by the

Bianchi theorem[42]. In accordance with the formula (5.2), the torus here is represented as a surface obtained by right translation of the asymptotic line $\gamma_1 = a(\cos u, \sin u, \cos u, \sin u)$ lying on it; with this representation, it proves that the second line $\Gamma = (\cos v, \sin v, 0, 0)$ does not lie on the surface (e.g., the point $(1, 0, 0, 0)$ belongs to this curve, but does not belong to the surface). If we would want to represent the torus as a translation surface of another asymptotic, the equation should be written as

$$(\cos u, \sin u, 0, 0) \cdot a(\cos v, \sin v, \cos v, -\sin v);$$

then the curve $\gamma_2 = a(\cos v, \sin v, \cos v, -\sin u)$, lying on the surface and obtained from Γ by left translation under the action of $\gamma_1(0)$, will be transferred. Thus, if the surface in the general case is already given as $r(u, v) = \gamma_1(u) \cdot \gamma_2(v)$, then the asymptotic lines lying on it and intersecting at the given point $M_0 = r(u_0, v_0)$ will be $\Gamma_1(u) = \gamma_1(u) \cdot \gamma_2(v_0)$ and $\Gamma_2(v) = \gamma_1(u_0) \cdot \gamma_2(v)$, and as the surface of transfer of its "base" asymptotic curves the surface $r(u, v)$ will be represented as $\Gamma_1 \cdot (\gamma_2(v_0)^{-1} \gamma_2(v))$ (the right translation of the "base" curve Γ_1) or as $(\gamma_1 \cdot \gamma_1(u_0)^{-1}) \cdot \Gamma_2$ (the left translation of the second "base" asymptotic curve). Therefore, we will never be able to write down the equation of the surface as $\gamma_1(u) \cdot \gamma_2(v)$ such that both "multipliable" curves γ_1 and γ_2 belong to the surface, excluding the case when there is a point $M_0(u_0, v_0)$ on it, in which $\gamma_1(u_0) = \gamma_2(v_0) = 1$. By the way, if by a respective choice of the coordinate axes it is achieved that the point $(1, 0, 0, 0)$ belongs to the surface, the Clifford torus can be written down in a more simple form as $(\cos u, \sin u, 0, 0) \cdot (\cos v, 0, 0, \sin v)$.

As we have already said above, a ruled Clifford surface with a flat metric can be constructed based on any curve in S^3. We will show that, proceeding from this observation, it is possible to obtain one necessary condition of the location of a curve in S^3. Let $\gamma : (y_1(u), y_2(u), y_3(u), y_4(u))$ be an arbitrary C^2-smooth curve in S^3

---

[42] Though on a Clifford torus for all cylindrical helices their constant torsion by modulus is less than 1 (there are non-closed ones among them, which give the dense cable of torus everywhere), but there are other examples of tori with base non-orthogonal circumferences, and for them we can explicitly write out two cylindrical helices with torsion ± 1, not being the asymptotic lines of the surface.

with the natural parameter u. Consider the surface

$$S : (y_1(u), y_2(u), y_3(u), y_4(u)) \cdot (\cos v, \sin v, 0, 0),$$

which, according to the above, has a flat metric

$$ds^2 = du^2 + 2f(u)\,dudv + dv^2, \quad f(u) = y_1 y_2' - y_2 y_1' - y_3 y_4' + y_4 y_3'. \quad (5.3)$$

Calculating the second form of the surface in coordinates (u, v), we obtain

$$II = Ldu^2 + 2Mdudv,$$

where

$$L = \frac{y_1'(y_2 + y_2'') - y_2'(y_1 + y_1'') - y_3'(y_4 + y_4'') + y_4'(y_3 + y_3'')}{\sqrt{1 - f^2}} \equiv$$

$$\frac{L_0}{\sqrt{1 - f^2}}, M = \sqrt{1 - f^2} \neq 0. \quad (5.4)$$

It is seen that even if the curve γ has a torsion ± 1, it does not have to be asymptotic, because for this we need the equality $L_0 = 0$. Let us substitute the coordinates such that the new coordinates be asymptotic. For this, it should be assumed that

$$u = C\xi, v = \eta + g(C\xi), g(u) = -\int \frac{L_0}{2(1 - f^2)}\,du,$$

where for the correctness of the required relations

$$I^2 = ds^2 = d\xi^2 + 2\cos\omega\,d\xi d\eta + d\eta^2, II = 2\sin\omega\,d\xi d\eta$$

the constant C should satisfy the condition

$$C^2 \left(1 - \frac{fL_0}{1 - f^2} + \frac{L_0^2}{4(1 - f^2)^2}\right) = 1. \quad (5.5)$$

As on the surface with an l. E. metric the transition to asymptotic coordinates is always possible, we obtain that the coordinates of any C^2-smooth curve in S^3 of necessity satisfy the condition

$$1 - \frac{fL_0}{1 - f^2} + \frac{L_0^2}{4(1 - f^2)^2} \equiv const > 0,$$

where the values of f and L_0 are determined in the formulae (5.3) and (5.4). Note that the expression for L_0 can be written down through the curvature and components of the principal normal of the curve. Unexpected here is the fact that in (5.5) only the second derivatives of the coordinates of the curve are involved.

Open question 5.2. Is there any relation between the expression (5.5) and the equation relating the curvatures of the curve on $S^3 \subset E^4$? (This equation was written out, for example, in [8].)•

Now we will show that the representability of a surface in the form of (5.2), where the curves γ_i have, respectively, torsions $\varkappa_i = (-1)^{i-1}, i = 1, 2$, is not only necessary, but is also sufficient for this surface to have an l. E. metric. This fact, certainly, is well known (see, for example, [154] and Chapter 7 in [166]), but we will give here the proof using only elementary information from differential geometry and close to papers by Bianchi [17, 18] and to Chapter V in [124], with consideration of S^3 as a unit sphere in the space of E^4 with the standard basis $e_1 = (1, 0, 0, 0), e_2 = (0, 1, 0, 0), e_3 = (0, 0, 1, 0), e_4 = (0, 0, 0, 1)$ and with the centre of the sphere in the origin of the coordinates.

Let $\gamma(u) = (x_1(u), x_2(u), x_3(u), x_4(u))$ be the natural equation of a curve in S^3 with torsion $\varkappa(u) = 1$ and curvature $k(u) \neq 0$. Let $t(u), n(u), b(u)$ be unit orts of the tangent, the principal normal and the binormal of the curve, with such a choice of their directions that the orientation of the four vectors γ, t, n, b coincide with the orientation of the standard basis. The Frenet equations for the curve γ have the following form:

$$\gamma'(u) = t, t'(u) = kn(u) - \gamma(u), n'(u) = -kt(u) + b(u), b'(u) = -n(u).$$
$$(5.6)$$

Besides, these vectors are related one with the other by the formula

$$b = (\gamma\gamma'n),$$
$$(5.7)$$

where the vector $(\gamma\gamma'n)$ has an i-th component equal to the minor (multiplied to $(-1)^i$) of the matrix

$$\begin{pmatrix} x_1 & x_2 & x_3 & x_4 \\ x_1' & x_2' & x_3' & x_4' \\ n_1 & n_2 & n_3 & n_4 \end{pmatrix},$$

obtained in cancelling its i-th column[43]. With account for the Frenet formulae, we have

$$b' = (\gamma\gamma''n) + (\gamma\gamma'n') = (\gamma\gamma'b),$$

which agrees with the last of the Frenet formulae, because for any orthonormalized system of vectors a, b, c, d from the equality $d = (abc)$ we have $a = -(bcd)$, $b = (cda)$ and $c = -(dab) = -(abd)$.

We obtained a linear equation for the vector $b(u)$, which, with account for the equalities $(b\gamma) = (b\gamma') = 0$, can be written down in the matrix form

$$b'(u) = M(u)b, \tag{5.8}$$

where $M = X'X^{-1}$, and

$$X = \begin{pmatrix} x_1 & -x_2 & -x_3 & -x_4 \\ x_2 & x_1 & -x_4 & x_3 \\ x_3 & x_4 & x_1 & -x_2 \\ x_4 & -x_3 & x_2 & x_1 \end{pmatrix}.$$

It is evident that the equation (5.8) has a solution of the form $b(u) = X(u)C$, where C is some constant matrix of the size (4×4). From the condition that at some initial value $u = u_0$ it should be $b(u_0) = X(u_0)C$ we obtain that $C = X(u_0)^{-1}b(u_0)$. Thus, the binormals to the curve γ with the torsion $\varkappa = 1$ are related to the initial direction of the binormal as

$$b(u) = X(u)X(u_0)^{-1}b(u_0),$$

which can be represented as the translation defined by the left action of the element $\gamma(u) \cdot \gamma(u_0)^{-1}$ on the initial binormal:

$$b(u) = (\gamma(u) \cdot \gamma(u_0)^{-1}) \cdot b(u_0). \tag{5.9}$$

Analogously, for curves with torsion $\varkappa = -1$ it could be shown that the binormal $b(u)$ is obtained by the right action on the binormal at the initial point:

$$b(u) = b(u_0) \cdot (\gamma(u_0)^{-1} \cdot \gamma(u)). \tag{5.10}$$

---

[43] This corresponds to the vector product of the vectors γ' and n in the space tangent to S^3 at the point $x = (x_1, x_2, x_3, x_4)$.

Let us turn now to surfaces given by the equation of the form (5.2) and consider for definiteness the surface S with the equation $r(u,v) = y(u) \cdot z(v)$. Let the lines $\Gamma_1(u) : r(u, v_0)$ and $\Gamma_2(v) : r(u_0, v)$ on the surface have torsions, respectively, $+1$ and -1. At the initial point $M_0 = r(u_0, v_0)$ they, by condition, have one common tangent plane, therefore, their binormals b_1 and b_2 are collinear at this point. As both these binormals are orthogonal to r'_u and r'_v, the normal ν to the surface is also collinear to them. The line $\gamma_1 : r = r(u_1, v) = y(u_1) \cdot z(v)$ passing through the point $M_1 = r(u_1, v_0)$ is obtained by the translation of the line $\Gamma_2 = y(u_0) \cdot z(v)$ by the left-hand-side action of the element $(y(u_1) \cdot y(u_0)^{-1})$ on it; by the formula (5.9), this element simultaneously also transfers the binormal to the line Γ_1 from the point M_0 to the point M_1 (keeping it as the binormal to the same line). As in the translation of the curve Γ_2 its binormal is also transferred, we have that at the point M_1 the curves Γ_1 and γ_1 have collinear binormals, therefore, along all the curve Γ_1 the normal to the surface is collinear to the binormal to this curve. A similar property is also valid along the curve Γ_2, and in the end over all the surface S the binormals to u- and v-lines prove collinear to the normal to the surface. Let us make use of this property to establish the local Euclidity of the metric of the surface S and to determine the form of the function $\omega(u,v)$ in the metric $ds^2 = du^2 + 2\cos\omega\,dudv + dv^2$ via the geometric characteristics of the curves Γ_1 and Γ_2. Let us write out for the surface the expressions of the second derivatives of its position vector (see Chapter V in [124]):

$$r_{uu} + r = \frac{\cos\omega}{\sin\omega}\omega_u r_u - \frac{\omega_u}{\sin\omega}r_v + L\nu$$

$$r_{uv} = -(\cos\omega)r + M\nu \qquad (5.11)$$

$$r_{vv} + r = -\frac{\omega_v}{\sin\omega}r_u + \frac{\cos\omega}{\sin\omega}\omega_v r_v + N\nu,$$

where ν is a unit normal to the surface, determined by the formula $\nu = (rr_u r_v)$. As the normal to the surface is collinear to the binormals to coordinate lines, then, by multiplying the first and third equations from (5.11) by ν, we obtain that $L = N = 0$. Then from the Peterson–Codazzi equations we obtain that $M = c\sin\omega$, and from the first equation of the system (5.11) with account for the Frenet equations we have:

$$(r_{uu} + r)^2 = k_1^2 = \omega_u^2,$$

where k_1 is the curvature of the curve Γ_1; similarly, from the third equations we obtain $\omega_v^2 = k_2^2$. Further, we have $(r_{uu}+r)(r_{uuv}+r_v) = \omega_u \omega_{uv}$ and, using the second equation of the system (5.11) and the Weingarten equation for expressing $\nu_u = $ via r_u and r_v it is already easy to establish the equality $(r_{uu} + r)(r_{uuv} + r_v) = 0$, i.e., $\omega_{uv} = 0$, which gives the local Euclidity of the metric of the surface. In passing, it can be shown that in the expression $M = c\sin\omega$ the coefficient $c = \pm 1$.

Further, let us choose the binormals to both asymptotic lines as co-directed with the normal ν to the surface, which, in turn, is chosen in the standard way as $\nu = a(rr_u r_v)$, where the coefficient $a > 0$ is selected such that the normal be singular. For this, take into account that the scalar square of the vector $(rr_u r_v)$ is equal to the Gram determinant of the vectors r, r_u, r_v; whence we obtain that $a = 1/\sin\omega$. Denote as n_1 the principal normal to the coordinate u line of Γ_1. From the Frenet equations we obtain $k_1 = (n_1 r_{uu})$. On the other hand, from (5.7) we have $n_1 = -(brr_u)$. Substituting the normal ν to the surface for the binormal b, from the equation $k_1 = a(rr_u r_u u)(rr_u r_v)$ with account for the first equation in (5.11) we obtain that (we simultaneously write the formula for k_2, too, because it is obtained by similar reasoning)

$$k_1 = -\omega_u', \quad k_2 = \omega_v'. \tag{5.12}$$

We can see that the choice made of the binormals' directions led to the situation that the curvatures of the asymptotic lines involved in the Frenet formula proved to be their geodesic curvatures, so their sign can be any, and, in particular, they can have vanishing points.

Thus, the surface with an l. E. metric in S^3 is formed by two families of congruent asymptotic lines with torsions ± 1; what is more, each of the families consists of lines obtained one from another by right (left) translations (shifts?) in the sphere S^3, considered as a Lie group of unit quaternions. Tangents to the asymptotic lines of one family in points of their intersection with the fixed asymptotic line of another family are left- or, respectively, right parallel in the Clifford sense [18].

Open question 5.3. Let a surface $r(u, v) = y(u) \cdot z(v)$ with a flat metric be given under standard assumptions for the curves (i.e., either both are circumferences or one of them is a circumference and the other is a curve with a corresponding torsion ± 1, or both have a torsion ± 1 of corresponding sign). Consider the surface $R(u, v) = z(v) \cdot y(u)$. In the case when one of the curves is a circumference, we know of the conditions when both surfaces have an l. E. metric. Then, if they are isometric, they are congruent. But in the general case the situation is more complicated. Examples show that these two surfaces can sometimes be congruent, that it can be that a surface of positive curvature with singular points is obtained, and it happens that the second surface in reality degenerates into a curve. Are there some sufficiently general criteria, which would enable any propositions on the surface $R(u, v)$? The question can be expanded thus: having a surface $y(u) \cdot z(v)$ with a flat metric, consider all combinations of products $y(u) \cdot y(v)$, $z(u) \cdot z(v)$, $z(v) \cdot y(u)$ or in a broad sense products of two different curves with the same torsion – both with the torsion $+1$ or both with the torsion -1. ●

Note that as a hypersurface with a flat metric in E^4 always has a ruled structure with plane two-dimensional generatrices [76], the two-dimensional surface in S^3 with a flat metric as a surface in E^4 cannot be a composition.

5.3   *Isometric immersions of locally Euclidean metrics into E^4 in the large*

In our view, the subject of this Section belongs to one of the very beautiful and intriguing problems of the geometry in the large, actively discussed in the literature over the last 15–20 years[44].

5.3.1   *Some general results*

In view of the vast literature, we will only briefly describe the main results available, the more so that their main part has already been

---

[44] Isometric embeddings or immersions of zero-curvature compact surfaces in E^3 are possible only in the class of C^1-smooth surfaces [103] or in the class of polyhedra with sufficiently fine triangulation [33].

dealt with in [25, 26], and the recent results are presented in a brief but capacious way in the Introduction to [5].

1) Let a closed asymptotic γ be on a C^3-smooth surface with an l. E. metric in S^3. Along the line γ, the angle ω between it and the asymptotic lines of another family intersecting this line is a periodic function (because $0 < \omega < \pi$ and the tangent to γ is periodic). Along γ, the function ω is a function of the arc length, so its derivative with respect to the arc is also periodic; therefore, the curve equal to this derivative is also periodic. The torsion, as a constant, is also periodic; then, by the Frenet formulae (5.6) we can see that the principal normal and the binormal are also periodic. And as along the asymptotic its binormal is collinear with the normal to the surface, then during the walk along the closed asymptotic the normal always returns to the initial position. Conclusion: *among C^3-smooth surfaces with an l. E. metric, in S^3 there exist no non-orientable surfaces with a closed asymptotic.*

Open question 5.4. When deducing the equalities (5.12), we used several formulae for calculating the curvature of an asymptotic line, the proofs of which formally used the C^3-smoothness of the surface. Is the now adduced proposition also true in the smoothness of class C^2?•

Open question 5.5. We can see that if a closed curve is the asymptotic line on some surface with a flat metric, then its Frenet trihedron will be periodic. In S^3, do closed semi-periodic curves exist, in particular, with constant torsion? (Recall that semi-periodicity means the periodicity of the tangent and antiperiodicity of the binormal and the principal normal.)•

2) By exactly the same reasoning, as we proved in Subsubsection 4.3.3.3 the impossibility of the existence of the analytical cylindrical Möbius surface, we can now prove that the previous proposition on the impossibility of a non-orientable zero-curvature surface in S^3 *also remains valid without the condition of the availability of a closed asymptotic line on the surface, but under the additional assumption of an analytical surface.* The idea of reducing the proof to the method of Subsubsection 4.3.3.3 is based on the fact that any analytical surface with an l. E. metric consists of a family of asymptotics translated

from two "base" asymptotics analytically extended maximally from their arcs on the surface considered; therefore, the normal to the surface is the binormal translated from them and during the walk along any closed curve it cannot come to the opposite direction.

Earlier, this proposition was established in [61], and in the class of smoothness C^∞ the impossibility of the existence of non-orientable surfaces with a flat metric in S^3 was proved in [97] at an additional assumption of the completeness of the surface. In [61] there is an example, which shows that in the class of C^∞ without the completeness condition the proposition is not true. The example is based on that in a non-analytical class the extension of an asymptotic line with the preservation of its torsion is not unique, and makes use of the idea of constructing a cylindrical Möbius band as in [42] described in detail in Subsubsection 4.3.3.3.

As the result, we have that in S^3 the topological types of complete surfaces with an l. E. metric can be only a plane, a cylinder and a torus.

5.3.2 *Isometric immersions and embeddings of a flat torus into E^4*

First we consider some concrete examples. Certainly, a trivial example of an embedding of a torus into E^4 is the direct product of two closed curves. In particular, the classical isometric immersion into R^4 of a torus abstractly given in the form of a rectangle with the sides a and b with respective identification of the opposite sides, known as the *generalized Clifford torus*, is given by the following formulae as the Cartesian product of two circumferences

$$x_1 = \frac{a}{2\pi}\cos\frac{2\pi u}{a}, \quad x_2 = \frac{a}{2\pi}\sin\frac{2\pi u}{a},$$

$$x_3 = \frac{b}{2\pi}\cos\frac{2\pi v}{b}, \quad x_4 = \frac{b}{2\pi}\sin\frac{2\pi v}{b}$$

(here $0 \le u \le a, 0 \le v \le b$). The obtained surfaces will in fact lie in three-dimensional spheres $S^3(r) \subset R^4$ of radius $r = \sqrt{a^2 + b^2}/2\pi$. Provided that $\sqrt{a^2 + b^2} = 2\pi$, i.e., when the internal diameter of the torus represented in the form of a rectangular lattice is equal to π, the embedding is into the unit sphere S^3 (in many works, exactly such a torus is called the Clifford torus). Apparently, the inverse is also true: the Cartesian product of two circumferences of radii $r_1^2 = \cos^2 \theta$ and $r_2^2 = \sin^2 \theta$ as a Riemannian manifold is represented in the form of a torus with the rectangular lattice with the standard metric of diameter π.

In this connection, we can immediately put the following question:

Open question 5.6. What tori with a flat metric can be isometrically embedded or immersed into the unit sphere S^3? ●

A partial answer to this question is known: if the internal diameter of the torus is not greater than π and the immersion of the torus into S^3 has the external (in the metric S^3) diameter π, then this torus can metrically be only a Clifford torus [57]; in particular, this is true if the torus is embedded. Below we formulate the proposition that no flat torus immersed into S^3 can be situated in a closed half sphere, therefore, any torus isometrically immersed into S^3 should have the internal diameter greater than $\pi/2$.

It turns out that it is possible to completely describe all immersions of tori into S^3 with the condition of constant angle ω between the asymptotic lines. Then circumferences will be its "base" asymptotics. Without loss of generality we can assume that the surface contains the point $(1, 0, 0, 0)$ and that one of the "base" circumferences is located on the plane $x_3 = x_4 = 0$. Then with accuracy of up to a motion all such surfaces have the form

$$x_1 = \cos \xi \cos \eta - \cos \omega \sin \xi \sin \eta$$

$$x_2 = \cos \xi \sin \eta + \cos \omega \sin \xi \cos \eta \qquad (5.13)$$

$$x_3 = \sin \omega \sin \xi \cos \eta, \quad x_4 = -\sin \omega \sin \xi \sin \eta,$$

where (ξ, η) are asymptotic coordinates, in which the metric $ds^2 = d\xi^2 + 2\cos \omega d\xi d\eta + d\eta^2$. Of greater interest is the description of those parallelograms with the standard metric $ds^2 = dx^2 + dy^2$, which admit such an isometric immersion into S^3. At a constant network angle ω the relation between the standard coordinates (x, y) and asymptotic coordinates (ξ, η) is expressed by a linear mapping[45] $x = \xi + \eta \cos \omega$, $y = \eta \sin \omega$, therefore, the domain of change of the

---

[45] In the general case this mapping is defined by a solution of the elliptic system $\eta_x + \cos \omega \xi_x + \sin \omega \xi_y = 0$, $\eta_y - \sin \omega \xi_x + \cos \omega \xi_y = 0$ with additional conditions

asymptotic coordinates will also be some parallelogram. Consider several possible cases.

1) A parallelogram on the plane (ξ, η) is a rectangle $0 \leq \xi \leq 2\pi m, 0 \leq \eta \leq 2\pi n$, where m and n are some natural numbers (this choice is dictated by the condition that the asymptotics $\xi = const$ and $\eta = const$ are circumferences). This case on the plane (x, y) is assigned parallelograms, two defining vertices of which $O(0,0)$ and $A(2\pi m, 0)$ are situated on the axis Ox, and the third is situated on the ray inclined at an angle ω to the side OA, at a distance of $2\pi n$ from the vertex O. Thus, in the considered class of surfaces an isometric immersion into S^3 is admitted by all parallelograms with the vertex in the origin of the coordinates and with two other vertices at points of the form $A(2\pi m, 0), m = 1, 2, ...$ and on the semi-circumferences of radii $2\pi n, n = 1, 2, ..$ with $y > 0$ (in particular, this set is assigned all rhombi with the side $2\pi m$). Asymptotic straight lines on the surface are assigned segments inside the parallelogram parallel to its sides and joining the identified points.

2) A parallelogram on the plane (ξ, η) is given by the inequalities $0 \leq \eta \leq l, 0 \leq \xi - q\eta \leq 2\pi m$. Then it could be shown that the parameters q and l should have the following representation:

$$l = (p + 2k)\pi, \quad q = \frac{p}{p + 2k},$$

where p and k are arbitrary integers with the condition $p + 2k > 0$. The indicated class of immersions is satisfied by all parallelograms with three vertices $O(0,0)$, $A(2\pi m, 0)$ and $B(l(q + \cos\omega), l\sin\omega)$ determining them. To obtain an explicit equation of their isometric immersion into S^3, it is required to substitute into (5.13) the values $\xi = x - y\,\cot\omega, \eta = \dfrac{1}{\sin\omega}y$. If immersions into spheres of arbitrary radius are admitted, then we obtain that the set flat tori with a constant angle between the asymptotics, immersed into the spheres, make an everywhere dense subset in the set of all metric tori.

$\xi_x^2 + \xi_y^2 = \eta_x^2 + \eta_y^2 = 1/\sin\omega$. As the transition to the universal covering yields an elliptical system on all the plane, it would be of interest to try to use this system for studies of tori in the sphere, in particular, to prove the periodicity of the asymptotic lines on the torus.

3) Consider separately the case when an immersed parallelogram is the rectangle $0 \leq x \leq 2\pi m, 0 \leq y \leq L$. Then from the previous text we have that the conditions

$$q + \cos\omega = 0, \ l\sin\omega = L$$

should be satisfied. Denoting $p + 2k = a$ with the even value of the difference $a - p$, we obtain that there exist the values of L, for which we can choose *different pairs* of natural numbers a and p, such that $a - p = 2n > 0$, but $L = \pi\sqrt{a^2 - p^2}$ is constant. Therefore, for these pairs we have different values of $\cos\omega$ and thereby we obtain examples of noncongruent isometric immersions of the same metric torus given as a rectangle. The smallest rectangle having two noncongruent immersions has the parameters $0 \leq x \leq 2\pi, 0 \leq y \leq 4\pi$. One of its immersions has the network angle $\omega = \pi/2$, and for its other immersion $\cos\omega = -3/5$.

Analogous noncongruent immersions of tori can also be found for other parallelograms.

But in a general case, when a flat torus is given as an arbitrary parallelogram ("skewed" torus), the solution of the problem of its isometric realization in E^4 proves a not altogether simple matter. As reported in [166], B. Lawson noticed that in the Hopf bundle $\pi : S^3 \to S^2$ of a unit sphere S^3 the curves from S^2 have in S^3 the pre-images, which are surfaces of zero curvature. As the pre-image $\pi^{-1}(p)$ of any point $p \in S^2$ is in S^3 a circumference, each closed curve c from S^2 determines in S^3 some flat torus $\pi^{-1}(c)$ (possibly, with self-intersections), and such tori in a paper by Pincall [122] are called Hopf tori (in the terminology used in the works by Bianchi [17, 18], they correspond to Clifford ruled tori; on them, one base asymptotic is a circumference; the other, a closed curve with torsion ± 1). From the results of this paper, we can have the following:

Theorem 5.1. *Each flat torus can be isometrically embedded into E^4 as the intersection of some algebraic surface of degree ≤ 4 with a three-dimensional sphere of some radius R.*

Indeed, each class of conformally equivalent tori has its unique "representative" among the set of flat tori T_0 defined by the parallelogram with vertices $(0, 0), (2\pi, 0), (a, b), (a + 2\pi, b)$, where $0 \leq a \leq$

$\pi, a^2 + b^2 \geq 4\pi^2$. Consider an aggregate of parallelograms T_1 defining some arbitrary class of flat tori isometric between themselves, without account for their orientation (i.e., only the isometricity of the parallelograms as tori is important, and the orientations of the parallelograms in establishing their isometry can be different). Using if necessary a homothetic transformation, let us change the aggregate of parallelograms T_1 such that the smallest length of the sides of the parallelograms become equal to 2π. Then in the set of parallelograms T_0 exactly one parallelogram from the changed aggregate of parallelograms T_1 will be found. And the work [122] has shown that each torus from the set T_0 admits an isometric embedding into S^3. Therefore, flat tori from T_1 can be isometrically embedded either into the unit sphere S^3 or into the sphere of some radius $R \neq 1$.

The method of constructing zero-curvature surfaces with account for their structure described in Subsections 4.4.2 and 4.4.3 using curves with torsion $\varkappa = \pm 1$ is far from giving the overall answer to the problem of the classification of all plane tori, mentioned in [196], p. 87 as a very interesting problem. First of all, it is not clear by which criteria we should take the "base" asymptotics so that the sine of the angle between them (which was different from zero in the beginning) remains different from zero in their extension (it is not difficult to construct examples, when this angle tends to 0 or to π). Moreover, it is necessary to try and get the surface obtained to be compact. The first step in this direction was made in [97], which proved that asymptotics on a torus are obliged to be closed lines (*a priori* they could be not closed); the other proofs of this fact are given in [137] and in [45]. What is more, [97] gives a method reducing the problem of finding the required pair of asymptotics in S^3 to finding the so-called "admissible pair" of closed curves γ_1 and γ_2 on the sphere $S^2 \subset S^3$, the main condition for which is the absence of intersection of the ranges of values of their curvatures. Over each curve a Hopf torus can be constructed, and on each torus a closed asymptotic taken; we denote them, respectively, Γ_1 and Γ_2. Then the surface $\Gamma_1 \cdot \Gamma_2^{-1}$ at a respective distribution of the signs of torsion of the curves $\Gamma_i, i = 1, 2$, will prove to be a torus with a flat metric, and in this way all tori are obtained.

Another approach to the description of flat tori in S^3, for example, in [55, 56, 187], makes use of their Grassmannian image.

One of the interesting properties of the torus in S^3 is the established ([98]) antipodal symmetry of the embedded torus, the proof of which requires very fine techniques with attraction of topological methods (see also [45]). In contrast, the following theorem from [147] is obtained by simple consideration of the extremum problem:

Theorem 5.2. *The isometric immersion of any flat torus into S^3 cannot be situated in a closed half sphere.*

Admit the inverse: some immersion of a torus is situated in a half sphere, which can be considered to be given by the inequality $x_1 \geq 0$. We introduce the function $x_1 = f = (er)$, where the vector $e = (1, 0, 0, 0)$, and r is the position vector of the surface. As the torus is a compact manifold, there exists a point of global minimum of the function f on it, in which in the local chart we have $f_u = (er_u) = 0$, $f_v = (er_v) = 0$. At this point the equality $(en) = 0$, where n is the normal to the surface, is not possible. Indeed, if $(en) = 0$, then at this point $r = \pm e$. The equality $r = e$ means that at this point f has a maximum equal to 1; the equality $r = -e$ means that then $x_1 = f = -1$, which contradicts the condition $x_1 \geq 0$. Hence, $(en) \neq 0$. We suppose that in the considered chart the metric has the standard form $ds^2 = du^2 + dv^2$. Then from the derivation formulae we obtain

$$f_{uu} = (er_u u) = -f + L(en), \quad f_{uv} = M(en), \quad f_{vv} = -f + N(en).$$

At the minimum point, we have

$$f_{uu}f_{vv} - f_{uv}^2 = f^2 - (L+N)(en)f - (en)^2 \geq 0 \qquad (5.14)$$

(it is accounted for that $LN - M^2 = -1$). Besides, we have the inequalities

$$f_{uu} = -f + L(en) \geq 0, \quad f_{vv} = -f + N(en) \geq 0.$$

We add up these two inequalities and obtain

$$-2f + (L+N)(en) \geq 0.$$

Let at the minimum point $f > 0$ (due to the inequality (5.14) at the minimum point the equality $f = 0$ is impossible). By multiplying the latter inequality by $f > 0$ and adding up with (5.14), we obtain the impossible inequality $-f^2 - (en)^2 \geq 0$. The obtained contradiction proves that the minimum of the function f is negative.

This theorem is a partial answer to the question from [98] on the existence of a pair of antipodal points on any immersion of a torus into S^3.

The works [188] and [61] study immersions and embeddings of a flat torus into E^4 "in the large". The main aim of [61] is to disprove the hypothesis expressed in [26] that any isometric immersion of a flat torus into E^4 is either the product of two curves or a torus in S^3. For this, the authors of the paper develop a new method of constructing tori in E^4, based on their concept of a plane mapping of the surface into S^3 and on the constructing of such mappings via solutions of some system of differential equations, which, in particular, make it possible to single out surfaces in R^4, which, in fact, lie in S^3. As the refutation of the above mentioned Borisenko hypothesis, the authors prove that in E^4 there exists a large family of tori with a flat metric and with flat normal connection, which are not the Cartesian product of two curves and do not lie in any affine sphere. In elaboration of these results, the work [5] defines and solves an analogue of the Cauchy problem for the search in S^3 of a surface with a flat metric as a band along a given curve on the sphere; in particular, this problem is solved for finding Möbius bands in S^3.

Dadok and Sha [45] provide a description of all isometric embeddings of a flat torus into S^3 and hypothesize that the entire plane R^2 does not admit an isometric embedding into S^3. Hereto, it can be added that there is as yet no explicit proof and embeddability of *any* cylinder into S^3, although the same work [45] proves the existence of embedded cylinders (if a cylinder embedded into S^3 is metrically represented as the product of the circumference to the straight line, then the radius of the circumference should apparently be no less than 1). As a matter of fact, it seems to us that it would be of interest to study in the theory of surfaces with flat metrics in S^3 and E^4 the following:

Open question 5.7. How can those domains in the Euclidean plane, an isometric immersion of which are the surfaces studied, be described?•

The formula (5.1) gives some hope for possible approaches to this problem, at least for plane surfaces in S^3.

5.3.3   *Analytical complete flat Möbius surfaces in E^4*

As we know from Subsection 4.3.3.3, a metrically complete flat Möbius band could not be either embedded or even immersed into E^3. But in E^4 such surfaces already exist, namely, the work [20] gives an explicit formula of an isometric embedding into E^4 of a Möbius strip with a complete metric in the analytical class of surfaces. We write down the most general form of such embeddings S, obtained by the idea of [20]:

$$x_1 = \rho \cos\left(\frac{u}{2} + h(\rho)\right), \quad x_2 = \rho \sin\left(\frac{u}{2} + h(\rho)\right),$$

$$x_3 = \frac{\sqrt{4-\rho^2}}{2}\cos(u + H(\rho)), \quad x_4 = \frac{\sqrt{4-\rho^2}}{2}\sin(u + H(\rho)), \quad (5.15)$$

where $0 \le u \le 2\pi$, and $h(\rho)$ and $H(\rho)$ are analytical even functions defined in some interval $-R < \rho < R \le 2$. It is directly verified that the obtained surfaces are a homeomorphic image of the rectangle $[0, 2\pi] \times (-R, R)$ with identification of the points $(0, \rho)$ and $(2\pi, -\rho)$. By calculating the metric, we have

$$ds^2 = (1 + \Phi(\rho))d\rho^2 + (du + \Psi(\rho)d\rho)^2,$$

where

$$\Phi(\rho) = \frac{\rho^2}{4(4-\rho^2)} + \frac{\rho^2(4-\rho^2)}{16}(H' - 2h')^2,$$

$$\Psi(\rho) = \frac{1}{4}(4H' - (H' - 2h')\rho^2).$$

For the metric to be complete, it is necessary and sufficient that the integral

$$\int_0^R |(2 - \tau)(H'(\tau) - 2h'(\tau))|d\tau$$

be divergent. Let us assume that it is so. After the substitution of the variables

$$t = t(\rho) = \int_0^\rho \sqrt{1 + \Phi(\tau)}d\tau, \quad v = u + g(\rho), \quad g(\rho) = \int_0^\rho \Psi(\tau)d\tau,$$

it is seen that the surfaces S have the l. E. metric $ds^2 = dt^2 + dv^2$, where the new variables (t, v) change on the Euclidean plane (v, t) in an infinite strip between the curves $\Gamma_1 : t = t(\rho), \ v = g(\rho)$ and $\Gamma_2 : t = t(\rho), \ v = 2\pi + g(\rho)$, with identification of the points $(t(\rho), g(\rho))$ and $(t(-\rho), 2\pi + g(-\rho))$ at these curves; herewith, the mid-line $t = 0$ passes in E^4 into a circumference $x_1 = x_2 = 0, \ x_3^2 + x_4^2 = 1$. When choosing $g(\rho) = 0$ we obtain the isometric embedding of an infinite Möbius strip 2π wide with rectilinear boundaries, and at $g(\rho) \neq 0$ we have an isometric embedding of an infinite strip with the curvilinear congruent boundaries. In [20], for the classical case $g(\rho) = 0$, which occurs at $h'(\rho) = \dfrac{\rho^2 - 4}{2\rho^2} H'(\rho)$, the author took

$H(\rho) = \ln(R^2 - \rho^2) + \dfrac{R^2}{R^2 - \rho^2}$, and at this choice of the functions $h(\rho)$ and $H(\rho)$ the transition to the standard metric $dt^2 + dv^2$ in the rectilinear strip could not be written out explicitly. But at $R = 2$ we can take the functions

$$H(\rho) = \frac{1}{2\sqrt{4 - \rho^2}} + \frac{\sqrt{4 - \rho^2}}{8}, \quad h = \frac{\sqrt{4 - \rho^2}}{16},$$

for which the transition to the standard coordinates (t, v) can be found in an explicit form:

$$t = \frac{7}{8}\rho + \frac{1}{8}\ln\frac{2 - \rho}{2 + \rho}, v = u \, (-2 < \rho < 2).$$

Further, it is not difficult to make certain that at any choice of $g(\rho)$ the surfaces S lie in a three-dimensional ellipsoid with the equation[46]

$$x_1^2 + x_2^2 + 4(x_3^2 + x_4^2) = 4$$

and therefore there are no rectilinear generatrices on them. The surface proposed in [20] is not algebraic, but if we take $h = H$ and, at that, with the divergent integral $\displaystyle\int_0^R |H'(\tau)|\,d\tau$ (e.g., it is sufficient

---

[46] This fact is of interest to compare with the proposition proved in [97] that in the class of smoothness C^∞ in the sphere S^3 there exists no complete non-oriented immersed surface with a flat metric.

to take the function $H(\rho)$ as in the example by Blanuša), then for any such choice with the even function H we obtain an algebraic surface giving a non-orientable isometric embedding into E^4 of an infinite plane strip with the congruent *curvilinear* boundaries. Indeed, at $H = h$, from the equations of the surface we have

$$x_1x_3 + x_2x_4 = \frac{\rho\sqrt{4 - \rho^2}}{2}\cos\frac{u}{2},\ x_1x_4 - x_2x_3 = \frac{\rho\sqrt{4 - \rho^2}}{2}\sin\frac{u}{2}.$$
$$(5.16)$$

After this, from pairs of equations for (x_1, x_2) and (x_3, x_4) we obtain

$$\cos h = \frac{x_1}{\rho}\cos\frac{u}{2} + \frac{x_2}{\rho}\sin\frac{u}{2}$$

$$\cos h = \frac{2x_3}{\sqrt{4 - \rho^2}}\cos u + \frac{2x_4}{\sqrt{4 - \rho^2}}\sin u.$$

Equating these expressions for $\cos h$ and substituting into the obtained equality the values of $\cos(u/2)$ and $\sin(u/2)$ found in (5.16), we obtain that the coordinates (x_1, x_2, x_3, x_4) of the points of the surface satisfy, besides the equation of the ellipsoid, also some equation of degree 5.

Open question 5.8. The constructed example and the approach used for this lead to the following questions.

a) How to construct an algebraic isometric embedding of a complete *standard* Möbius strip into E^4?

b) If such surfaces are proved to exist, then what is their *lowest* order?

c) The method used for constructing the isometric embedding of a complete Möbius strip 2π wide into E^4 is also suitable for embedding into E^4 Möbius strips of any width a, but all of them will be situated in ellipsoids of different parameters depending on a. Do arbitrary Möbius bands admit embedding into E^4 such that all of them, irrespective of the width of the strip, be situated in one ellipsoid?•

Now we note that if in the formula (5.15) some functions $a(\rho)$ and $b(\rho)$ are substituted, respectively, for the coefficients ρ and $\sqrt{4 - \rho^2}/2$ at the cosines and sines, with the condition $a^2 + 4b^2 = 4$, then we obtain a greater freedom of search of the Möbius strip embedded into

R^4. Along with the freedom of choice of the functions $h(\rho)$ and $H(\rho)$ this means that the Blanuša method yields an embedding into the ellipsoid $x_1^2 + x_2^2 + 4x_3^2 + 4x_4^2 = 4$ of a complete Möbius strip in the form of nontrivially bendable surfaces with the freedom of choice of three functions of one variable (two functions if we embed a rectilinear strip).

5.3.4 Another isometric embedding of an infinite Möbius strip into E^4

Along with the above mentioned (see the footnote on p. 205) result by Kitagawa [97] there exists one more "negative" result by Weiner [187] on the impossibility of an isometric embedding of a flat Klein bottle into S^3 (and even into E^4 with an additional condition of triviality of normal connection of the surface and the regularity of its Gaussian mapping). In this Subsection, we will describe one more method of isometric embedding of a complete rectilinear Möbius strip $\Pi : -A \le s \le A, -\infty < t < \infty$ into E^4, which makes it possible to construct an isometric embedding into E^4 of a flat Klein bottle, too. This method is presented in [34] and is very geometric by its idea of construction. However, we should say that the work has a typographical error, which could be found only after making certain that the written formulae are wrong and choosing the value of one parameter required for them to be valid. For this reason, we present the construction of the sought-for surface in a corrected version with required refining, the more so that in [34] the construction is given in a very brief form.

We introduce into consideration the mapping $h : R^2 \to R^2$ defined by the formula

$$h(x) = (h_1(x), h_2(x)) = \begin{cases} (\tfrac{l}{2} + x, 0) & at \quad x \in (-\infty, -\tfrac{l}{2}], \\ (g_1(x), g_2(x)) & at \quad x \in [-\tfrac{l}{2}, \tfrac{l}{2}], \\ (\tfrac{l}{2} - x, 0) & at \quad x \in [\tfrac{l}{2}, +\infty), \end{cases}$$

where l is an arbitrary positive constant, and the functions $g_i(x), i = 1, 2$, are such that in the large $h(x) \in C^\infty$ and the component $h^1(x)$ is even, $h^2(x)$ is odd, and the argument x is a natural parameter of the curve $h(x)$. Besides, we will assume that $h_1(x)$ is monotone at

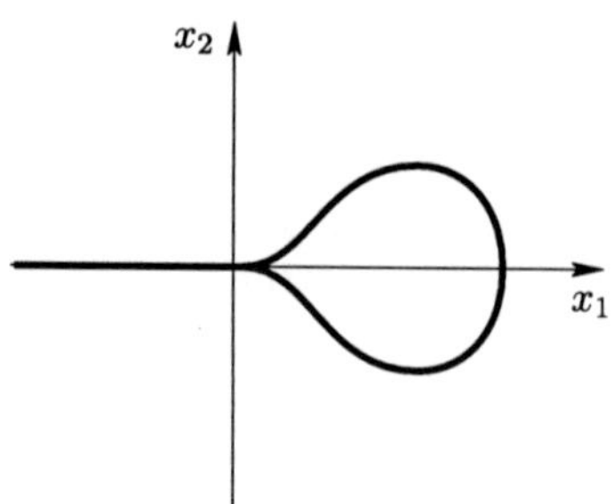

Figure 17 (courtesy of *Izvestiya: Mathematics*)

each segment $(-\infty, 0)$ and $(0, +\infty)$, and the entire curve $h(x)$ has a self-intersection only at points with $|x| \geq l/2$ (see Figure 17).

Now we are given a curve $f(t) = (f_1(t), f_2(t)), t \in [-a, a], 0 < a \leq \infty$, satisfying the conditions: $f(t)$ is of class C^∞, the parameter t for it is natural, the component $f_1(t)$ is odd and bounded and the component $f_2(t)$ is even. Consider the surface S_1 in E^4, given by the mapping $X_1(s,t) : \mathbb{R} \times [-a, a] \to R^4$ according to the formula

$$
X_1(s,t) = \begin{cases}
x_1 = & h_1(s \cos\theta + f_1(t) \sin\theta) \\
x_2 = & h_2(s \cos\theta + f_1(t) \sin\theta) \\
x_3 = & -s \sin\theta + f_1(t) \cos\theta \\
x_4 = f_2(t),
\end{cases} \tag{5.17}
$$

where the parameter $\theta \in (0, \pi/2)$, and the arguments $s \in \mathbb{R}, t \in [-a, a]$. By direct calculation, we verify that $dx_1^2 + dx_2^2 + dx_3^2 + dx_4^2 = ds^2 + dt^2$, so that the surface S_1 is an isometric immersion of the domain $-\infty < s < +\infty, \ -a \leq t \leq a$ into E^4; herewith, if the curve $f(t)$ is closed, then S_1 will be the image of a cylinder. Hereafter, we consider that the curve $f(t)$ is not closed and has no self-intersections, the bounded component f_1 satisfies the inequality $|f_1(t)| < (l/2) \sin\theta$ for all t, and the component $f_2(t)$ at each of the segments $(-\infty, 0)$ and $(0, +\infty)$ is monotone. Under these assumptions, we prove that the *surface S_1 is embedded* (in corresponding reasoning, [34] considered not all possibilities and, moreover, without our conditions for the curve $f(t)$ the proposition on the embeddability of S_1 may prove false).

Indeed, if at the points $M_1(s_1, t_1)$ and $M_2(s_2, t_2)$ the values of the mapping X_1 are equal, then the curve $h(x)$ consisting of the points $(x_1, x_2, 0, 0)$ has a self-intersection; therefore, at the self-intersection point $|s \cos \theta + f_1(t) \sin \theta| \geq l/2$. Consider two possible variants of the values of the argument in $h_1(x)$: $a)$ $s_1 \cos \theta + f_1(t_1) \sin \theta \geq l/2$, $s_2 \cos \theta + f_1(t_2) \sin \theta \geq l/2$ and $b)$ $s_1 \cos \theta + f_1(t_1) \sin \theta \geq l/2$, $s_2 \cos \theta + f_1(t_2) \sin \theta \leq -l/2$ (two more possible variants are studied similarly). In the first case, from the equalities of two values of h_1 we have $(s_1 - s_2) \cos \theta + (f_1(t_1) - f_1(t_2)) \sin \theta = 0$. We add here the equation

$$(s_1 - s_2) \sin \theta - (f_1(t_1) - f_1(t_2)) \cos \theta = 0, \qquad (5.18)$$

which follows from the equality of two values of x_3, and obtain $s_1 - s_2 = 0$, $f_1(t_1) - f_1(t_2) = 0$. The latter relation, along with $x_4 = f_2(t_1) = f_2(t_2)$ means that the nonclosed curve $f(t)$ has a self-intersection, which is excluded by condition; therefore, $s_1 = s_2, t_1 = t_2$. In the second case, with account of the form $h_1(x)$ at $x \leq -(l/2)$ and $x \geq l/2$ we obtain that $(s_1 + s_2) \cos \theta + (f_1(t_2) + f_1(t_1)) \sin \theta = 0$. Recalling the conditions for $f_2(t)$, from the equality of two values of x_4 we have that $t_2 = -t_1$. From the equality of two values of x_1 with account for the oddity of $f_1(t)$ we obtain $s_1 + s_2 = 0$, but then from (5.18) we have $s_1 \sin \theta - f_1(t_1) \cos \theta = 0$. From the restriction to $|f_1(t)|$ it follows that $|s_1| < (l/2) \cos \theta$. But in this case, we have $s_1 \cos \theta + f_1(t_1) \sin \theta < l/2$, $s_2 \cos \theta + f_1(t_2) \sin \theta = -s_1 \cos \theta - f_1(t_1) \sin \theta > -l/2$, which contradicts the considered variants. The proposition on the absence of self-intersection is proved.

Let us now use the surface S_1 to construct an embedding of an infinite Möbius strip into E^4 defined in the form of an infinite strip between two straight lines with respective identification of points of the boundary lines. Let S_1 be an embedding of the rectangle $\Pi_A : -A \leq s \leq A$, $-\infty = -a < t < a = +\infty$ into E^4 by the mapping (5.17), in which $\theta = \pi/6$ (in [34], the value $\pi/3$ was taken, which subsequently gives an erroneous proposition on the mapping $X_0(s, t)$). Let us show that the parameter A can be chosen so that the projection of the surface S_1 to the plane (x_1, x_3) lie between the rays emanating from the point $(l/2 - 2\sqrt{3}A/3, 0, 0, 0)$ in the direction of the vectors $e_1 = \{1, 0, \sqrt{3}, 0\}$ and $e_2 = \{1, 0, -\sqrt{3}, 0\}$ and forming an angle of $2\pi/3$ between themselves. For this, let us first find the images of the straight lines $L_1 : s = -A$ and $L_2 : s = A$ on the plane (x_1, x_3), assuming that the parameter A is chosen with the conditions $\sqrt{3}A + f_1(t) \geq l$ and $-\sqrt{3}A + f_1(t) \leq -l$ (these conditions give us the arguments in h_1, by their absolute value not smaller than $l/2$, and thus enabling the use of the explicit form of the function $h_1(x)$ at such values of the argument). Considering that we already have the requirement $|f_1(t)| < (l/2)\sin\theta = l/4$, we obtain that to simultaneously satisfy all conditions, it is sufficient to choose the values of A from the following range

$$\frac{4\sqrt{3}}{12}l < A < \frac{5\sqrt{3}}{12}l.$$

Under these conditions we have:

$$x_1(-A, t) = h_1\left(-\tfrac{\sqrt{3}}{2}A + \tfrac{1}{2}f_1(t)\right) = \tfrac{l}{2} - \tfrac{\sqrt{3}}{2}A + \tfrac{1}{2}f_1(t),$$
$$x_3(-A, t) = \tfrac{1}{2}A + \tfrac{\sqrt{3}}{2}f_1(t),$$

which gives us the equation of the straight line $l_1 : \sqrt{3}x_1 - x_3 + 2A - \frac{\sqrt{3}}{2}l$. Similarly, it can be shown that points of the straight line L_2 are mapped into points of the straight line $l_2 : \sqrt{3}x_1 + x_3 + 2A - \frac{\sqrt{3}}{2}l$. These straight lines intersect at the point $M_0(l/2 - 2\sqrt{3}A/3, 0)$ and split the plane (x_1, x_3) into four parts. Let us make certain that the respective surfaces S_1 lie inside an angle containing the origin of coordinates. Consider the functions

$$F_1(s, t) = \sqrt{3}x_1(s, t) - x_3(s, t) + 2A - \frac{\sqrt{3}}{2}l,$$

$$F_2(s, t) = \sqrt{3}x_1(s, t) + x_3(s, t) + 2A - \frac{\sqrt{3}}{2}l.$$

On the straight line $L_1 : s = -A$ the function $F_1(s, t) = 0$, on the straight line $L_2 : s = A$ we have $F_1(A, t) > \sqrt{3}l > 0$. At the points (s, t), where $C \equiv \frac{\sqrt{3}}{2}s + \frac{1}{2}f_1(t) \leq \frac{l}{2}$, we have $F_1(s, t) > 0$.

Indeed, $F_1(s, t) = \sqrt{3}h_1(C) + \frac{s}{2} + 2A - (l + f_1(t))\frac{\sqrt{3}}{2}$. In this expression the summands $\frac{s}{2} + 2A - (l + f_1(t))\frac{\sqrt{3}}{2}$ at the given restrictions to A and $f_1(t)$ always yield a positive sum. At the values of the argument, by module not exceeding $l/2$, the function $h_1(x)$ is non-negative, therefore, if $|C| \le l/2$, then $F_1(s,t) > 0$. But if $C < -(l/2)$, then we can calculate $F_1(s,t)$ by the definition of the function $h_1(C)$. We obtain

$$F_1(s,t) = \sqrt{3}\left(\frac{l}{2} + C\right) + \frac{s}{2} + 2A - (l + 2C - \sqrt{3}s)\frac{\sqrt{3}}{2} = 2A + 2s > 0.$$

At the straight line $s = A$ the values of $C = \dfrac{\sqrt{3}}{2}A + \dfrac{1}{2}f_1(t) > \dfrac{l}{2}$ (in reality, at $t = 0$ we have $C = \sqrt{3}A/2 > l/2$, and if somewhere $C = \dfrac{\sqrt{3}}{2}A + \dfrac{1}{2}f_1(t) = \dfrac{l}{2}$, we have a contradiction with the condition $f_1(t) > l - \sqrt{3}A$). At the straight line $s = -A$ the values of $C = -\dfrac{\sqrt{3}}{2}A + \dfrac{1}{2}f_1(t) < 0$, therefore, the graph Γ of the curve $\dfrac{\sqrt{3}}{2}s + \dfrac{1}{2}f_1(t) = \dfrac{l}{2}$ passes inside the band, nowhere intersecting its boundary lines, and having a single-valued projection on the vertical axis t. It splits all the band into two parts – on the left from Γ, where $F_1(s,t) > 0$, and on the right from Γ; we designate this "right" domain as D^+. In D^+, let us consider the segments $t = const$. They begin on the curve Γ and go to the right up to the straight line $s = A$. Along these segments the derivative $\dfrac{\partial F_1(s,t)}{\partial s} = \dfrac{3}{2}h_1'(C) + \dfrac{1}{2} = -1$ (because $h_1'(C) = -1$ at the argument $C > l/2$). Therefore, the function F_1 monotonously decays along these segments, and as at both ends of the segments it is positive, then $F_1(s,t) > 0$ in D^+, hence, it is positive everywhere in the band. And the positive values of the function F_1 as a function of (x_1, x_3) are situated in the same half-plane as the origin of coordinates.

In a similar way, it is proved that the function $F_2(s,t)$ is also positive everywhere in the band, and this sign on the plane (x_1, x_3) corresponds to the half-plane containing the origin of coordinates. As the result, we proved that in the mapping X_1 the band Π_A passes into the domain defined by the inequalities

$$\sqrt{3}x_1 - x_3 + 2A - \frac{\sqrt{3}}{2}l > 0, \ \sqrt{3}x_1 + x_3 + 2A - \frac{\sqrt{3}}{2}l > 0,$$

i.e., into the angle between the above mentioned rays.

Now we will construct two mappings $X_2(s,t)$ and $X_3(s,t)$ of the same band Π_A into E^4. These mappings will be assigned the surfaces S_2 and S_3 obtained from S_1 by rotating the plane (x_1, x_3) to the angle $\pm(2\pi/3)$ around the point $M_0(l/2 - 2\sqrt{3}A/3, 0)$, being the vertex of the angle $2\pi/3$, in which the domain for the mapping X_1 is situated. The corresponding formulae have the form:

$$\left\{ \begin{array}{ll} x_1= & -\frac{1}{2}h_1\left(s\sqrt{3}/2+f_1(t)/2\right)+s\sqrt{3}/4-3f_1(t)/4+3l/4-A\sqrt{3} \\ x_2= & h_2\left(s\sqrt{3}/2+f_1(t)/2\right) \\ x_3= & \frac{\sqrt{3}}{2}h_1\left(s\sqrt{3}/2+f_1(t)/2\right)+s/4-\sqrt{3}f_1(t)/4-l\sqrt{3}/4+A \\ x_4= & f_2(t), \end{array} \right.$$

for $X_2(s,t)$ and

$$\left\{ \begin{array}{ll} x_1= & -\frac{1}{2}h_1\left(s\sqrt{3}/2+f_1(t)/2\right)-s\sqrt{3}/4+3f_1(t)/4+3l/4-A\sqrt{3} \\ x_2= & h_2\left(s\sqrt{3}/2+f_1(t)/2\right) \\ x_3= & -\frac{\sqrt{3}}{2}h_1(s\sqrt{3}/2+f_1(t)/2)+s/4-\sqrt{3}f_1(t)/4+l\sqrt{3}/4-A \\ x_4= & f_2(t). \end{array} \right.$$

for $X_3(s,t)$.

As all the domain for $X_1(s,t)$ on the plane (x_1, x_3) is located inside the angle of $120°$ between two rays, after a turn to $\pm 120°$ the domains for $X_2(s,t)$ and $X_3(s,t)$ will not intersect, each being located in its sector (Figure 18), so the united surface $S_0 = S_1 \cup S_2 \cup S_3$ will have no intersections. By direct calculations, we find that $X_1(-A, -t) = X_2(A,t)$, $X_1(A,t) = X_3(-A,-t)$, $X_2(-A,-t) = X_3(A,t)$. These equalities make it possible to assert that the mapping $X_0(s,t)$ into E^4 of the band $\Pi : -3A \leq s \leq 3A, -\infty < t < +\infty$, defined by the following formulae:

$$X_0(s,t) = \left\{ \begin{array}{ll} X_2(2A+s,t), & -3A \leq s \leq -A \\ X_1(s,-t), & -A \leq s \leq A \\ X_3(-2A+3,t), & A \leq s \leq 3A \end{array} \right.$$

will be continuous in the large. What is more, on the left- and right-hand side of the lines of glueing $s = -A, s = A$ and $s = -3A, s = 3A$ it is given by the formulae, which are alike in their appearance. For

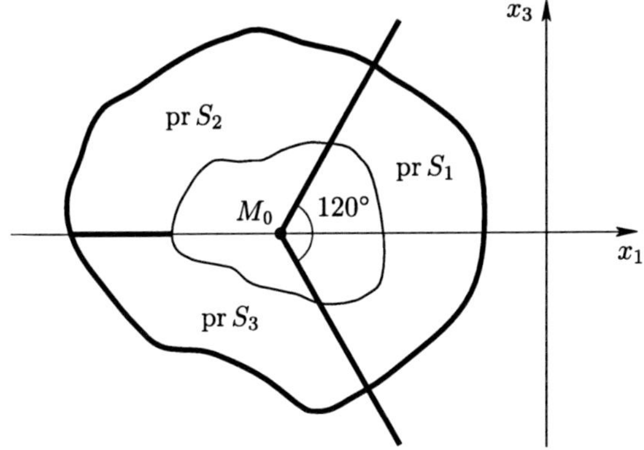

Figure 18 (courtesy of *Izvestiya: Mathematics*)

instance, in the neighbourhood of the line $s = -A$ it has the following representation:

$$X_0(x - A, t)(x < 0) = X_2(x + A, t):$$
$$\begin{cases} x_1 = & (x - A)\sqrt{3}/2 + l/2 - f(t)/2 \\ x_2 = & 0 \\ x_3 = & (A - x)/2 + A/2 - f(t)\sqrt{3}/2 \\ x_4 = & f_2(t) \end{cases}$$

and

$$X_0(x - A, t)(x > 0) = X_1(x - A, -t):$$
$$\begin{cases} x_1 = & (x - A)\sqrt{3}/2 + l/2 - f_1(t)/2 \\ x_2 = & 0 \\ x_3 = & (A - x)/2 + A/2 - f_t(t)\sqrt{3}/2 \\ x_4 = & f_2(t), \end{cases}$$

i.e., near the lines of glueing it has the same smoothness as the curve $f(t)$ (which can also be defined as an analytical curve). As for points of the boundary straight lines $s = -3A$, $s = 3A$ of the band Π on the surface S_0 identification occurs according to the law $X_0(-3A, -t) = X_0(3A, t)$ (and inside the band in passing via the lines of glueing the orientation does not change), the constructed surface is indeed an isometric embedding of an infinite Möbius strip into E^4; what is more, the smoothness of embedding coincides with the minimal smoothness of the curves $h(s)$ and $f(t)$. We also note that the curve $f(t)$ situated on the plane (x_1, x_4) (with a shift to $l/2 - A\sqrt{3}$ along the axis x_1) is the image of the boundary lines $s = -3A$ and $s = 3A$.

Remark 5.1. Though in [34] it is asserted that the method of this work is suitable also for constructing an isometric embedding of any "skewed" torus (i.e., a torus produced by identification of the sides of an arbitrary parallelogram) into E^4, this construction still awaits its realization.•

5.3.5 Isometric immersion of a Klein bottle into R^4

This immersion is given in [176] by the following mapping of the rectangle $0 \leq u \leq 2\pi, 0 \leq v \leq 2\pi$

$$x_1 = \cos u \cos v, \quad x_2 = \sin u \cos v,$$
$$x_3 = 2\cos\frac{u}{2}\sin v, \quad x_4 = 2\sin\frac{u}{2}\sin v,$$

with the l. E. metric $ds^2 = du^2 + (\sqrt{1 + 3\cos^2 v}\,dv)^2$ and by identification of the sides of the rectangle by the rule $(u, 0) \leftrightarrow (u, 2\pi), (0, v) \leftrightarrow (2\pi, 2\pi - v)$. We obtain an algebraic surface with the equations $4(x_1^2 + x_2^2) + x_3^2 + x_4^2 = 4$, $2x_1x_3x_4 - x_2x_3^2 + x_2x_4^2 = 0$ (note that again, as in Blanuša for the complete Möbius strip, in the same three-dimensional ellipsoid we have a non-oriented analytical surface with a flat metric, this time a compact surface). When finding a "natural" domain of standard parameters, we find that the rectangle $0 \leq u \leq 2\pi, 0 \leq t \leq T$ with the metric $ds^2 = du^2 + dt^2$, where $T = \int_0^{2\pi} \sqrt{1 + 3\cos^2 v}\,dv \approx 9,7$, is isometrically embedded.

We will show how *any* Klein bottle can be immersed into E^4. Let the metric of the Klein bottle is abstractly given as the metric of a rectangle $P_{cd} : 0 \leq x \leq c, 0 \leq y \leq d$ with the locally metric form $ds^2 = dx^2 + dy^2$, and the identification is performed with the preservation of the direction for the sides $y = 0$ and $y = d$ and with the opposite walk for the sides $x = 0$ and $x = c$. Consider the rectangle Π :

$0 \le u \le u_0$, $0 \le v \le 2\pi$, and its immersion into E^4 by the mapping

$$x_1 = a\cos(bu + H(v))\cos v, \quad x_2 = a\sin(bu + H(v))\cos v$$

$$x_3 = 2a\cos\left(\frac{bu}{2} + h(v)\right)\sin v, \quad x_4 = 2a\sin\left(\frac{bu}{2} + h(v)\right)\sin v, \quad (5.19)$$

where $H(v)$ and $h(v)$ are 2π-periodic even functions, and the coefficient b is chosen such that the equality

$$bu_0 = 2\pi(2n+1) \tag{5.20}$$

be satisfied with some non-negative integer n.

The coincidence of images in mapping (5.19) pairs of points $(u, 0)$, $(u, 2\pi)$ and $(0, v)$, $(u_0, 2\pi - v)$ is checked directly, so the obtained surface is topologically a Klein bottle. By calculating the coefficients of the metric form of the surface, we find

$$E = a^2 b^2, \quad F = ba^2(H'(v)\cos^2 v + 2h'(v)\sin^2 v),$$

$$G = a^2(\sin^2 v + 4\cos^2 v + H'^2(v)\cos^2 v + 4h'^2(v)\sin^2 v).$$

Then the metric form can be written down as

$$ds^2 = a^2\left(bdu + (H'(v)\cos^2 v + 2h'(v)\sin^2 v)dv\right)^2 +$$
$$a^2\left(1 + 3\cos^2 v + \Phi(v)\right)dv^2,$$

where $\Phi(v) = (2h' - H')^2 \cos^2 v \sin^2 v$. We choose the functions H and h such that the equation $H'(v)\cos^2 v + 2h'(v)\sin^2 v = 0$ be satisfied. For instance, let

$$H = -\alpha(\cos v - \frac{1}{3}\cos^3 v), \quad h = \frac{\alpha}{6}\cos^3 v.$$

Then the substitution of the variables

$$x = abu \text{ and } y = y(v) = a\int_0^v \sqrt{1 + 3\cos^2 \tau + \alpha^2 \cos^2 \tau \sin^4 \tau}\,d\tau$$

brings the metric of the surface to the standard form $ds^2 = dx^2 + dy^2$ over some rectangle. It remains to achieve that this rectangle have the necessary size. We want it to be

$$abu_0 = c \text{ and } af(2\pi, \alpha) = d,$$

where

$$f(2\pi, \alpha) = \int_0^{2\pi} \sqrt{1 + 3\cos^2\tau + \alpha^2 \cos^2\tau \sin^4\tau}\, d\tau.$$

With account for (5.20) we have

$$a = \frac{c}{2\pi(2n+1)}.$$

By choosing a sufficiently large n, we can get that $af(2\pi, 0) \leq d$. The function $f(2\pi, \alpha)$ grows monotonously and infinitely at the increase of α, so we can find such a value of $\alpha = \alpha(a)$, for which $af(2\pi, \alpha(a)) = d$. Therefore, we obtained an infinite discrete set of values of the three coefficients $b_n, a_n, \alpha(a_n), n \geq n_0$, where n_0 is the first integer with the condition $\dfrac{c}{2\pi(1 + 2n_0)} f(2\pi, 0) \leq d$, for each of which the mapping (5.19) after the transition to the variables (x, y) performs an isometric immersion into E^4 of the given rectangle P_{cd} in the form of a Klein bottle. By some complication we can, apparently, achieve that a mapping of the form (5.19) yield as the result algebraic surfaces lying, irrespective of the size of the immersed rectangle, in the ellipsoid $4(x_1^2 + x_2^2) + x_3^2 + x_4^2 = 4$. The impression is that this ellipsoid plays the same role for non-oriented surfaces with flat metrics as a unit sphere does for a flat torus. It can be shown that if the relation $d : a$ is sufficiently small, then the mapping (5.18) at the chosen $\varkappa(v)$ and $h(v)$ will be an embedding. Apparently, by a suitable choice of $\varkappa(v)$ and $h(v)$ we can succeed in the mapping by the formula (5.18) be an embedding at all c and d.

Remark 5.2. The second isometric immersion of the Klein bottle from [87],

$$x_1 = \cos u, \quad x_2 = \sin u, \quad x_3 = a\sin(v + \frac{u}{2}), \quad x_4 = b\sin 2(v + \frac{u}{2}),$$

mentioned in the literature (see [26, 69] and [128]), unfortunately proved wrong: this surface in reality is a torus. Indeed, the considered functions yield a mapping of the parallelogram $\Pi : (0 \leq u \leq 2\pi, u/2 \leq v \leq 2\pi + u/2)$ into R^3 with the following identification of the sides:

$$(u, v = u/2) \leftrightarrow (u, v = 2\pi + u/2), \quad (0, v) \leftrightarrow (2\pi, v + \pi), 0 \leq u, \ v \leq 2\pi,$$

i.e., this identification does not change the orientation (all other variants of coincidence of the images of points of the plane (u, v) also lead to mappings of some parallelograms without changing the orientation of no pairs of identified sides). If we substitute the coordinates $u = u, v + u/2 = V$, we arrive at the mapping

$$x_1 = \cos u, \quad x_2 = \sin u, \quad x_3 = a \sin V, \quad x_4 = b \sin 2V,$$

which appears as the product of two flat curves $\Gamma_1 : (\cos u, \sin u, 0, 0)$ and $\Gamma_2 : (0, 0, a \sin V, b \sin 2V)$, each of which goes without the change of direction of the walk.$\bullet$

In this way we obtained the confirmation of the proposition, presented in [66], 3.2.1(F), without the proof, on that *any* flat Klein bottle admits an isometric analytical immersion into E^4. But for a more general class of smoothness, namely, for class C^∞, the work [34] gives a method of an isometric *embedding* of an arbitrary flat Klein bottle into E^4. Recall that for the earlier constructed embedding of an infinite Möbius band we have noted that the edges of the band are mapped into E^4 as a curve $f(t)$. Therefore, if in the mapping $X_0(s, t)$ we take as $f(t)$ a closed curve without self-intersections with the previous conditions for the components and with the length $2L$, then at the mapping $X_0(s, t), -3A \leq s \leq 3A, -L \leq t \leq L$ in view of the equality $f(-L) = f(L)$ the points of the edges $t = -L, t = L$ of the rectangle Π_{3A} are glued according to the law $(s, -L) \leftrightarrow (s, L)$, with the satisfaction of the previous equalities $X_0(-3A, -t) = X_0(3A, t)$, i.e., the sides of the rectangle Π_{3A} will prove to be identified as in the topological model of the Klein bottle.

As for any Klein bottle[47] its metric model can be represented in the form of a rectangle with respective identification of the sides ([192], Section 2.5, therewith we have a C^∞-smooth isometric embedding of any Klein bottle into E^4.

Still unsolved are the following questions:

Open question 5.9. 1) Can it be affirmed that in any three-dimensional ellipsoid in E^4 there exists a Klein bottle with a flat metric?

---

[47] We mean that *a priori* any two-dimensional non-oriented manifold, having the Euler characteristic $\chi = 0$ and equipped with a locally Euclidean metric, *can be called* a flat Klein bottle.

In more general terms, is the analogue of Theorem 4.5 with the substitution of an ellipsoid for a sphere is valid for Klein bottles?

2) Can it be affirmed that all flat Klein bottles are isometrically embeddable into E^4 in the form of an algebraic surface of order 3?

3) Find an algebraic embedding of a complete flat Möbius strip, given in the standard way in the form of a band with the parallel rectilinear edges, into E^4.

4) Find any explicit formulae for an analytical and, in particular, algebraic or C^∞-smooth isometric embedding into E^4 or at least an immersion into E^4 of *any* flat "skewed" torus given in the form of a parallelogram with corresponding identification of its opposite sides. Recall that it is sufficient to be able to immerse tori in the form of a parallelogram with the vertices $(0,0), (2\pi, 0), (a, b), (a + 2\pi, b)$, where $0 < a \le \pi, a^2 + b^2 > 4\pi^2$, as the other tori are reduced to them by some elementary transformations. By the theorem from [122] all these tori admit an isometric embedding into S^3 of some radius in the form of an algebraic surface of degree not higher than 4, so at least some algorithm could be found for constructing such surfaces by the metric torus given in the form of a parallelogram.

5) The problem in a sense converse to the previous problem: how, by a flat torus in S^3 given in some way, can we determine the parallelogram, of which the torus is an isometric immersion?•

6 Surfaces with locally Euclidean metrics in multidimensional spaces

6.1 *Isometric immersions of n-dimensional locally Euclidean metrics into E^n*

We will be working with several copies of the space R^n, which for greater clarity we will designate as R_u^n or R_x^n in accordance with the designation of the coordinates in this space: $u_1, \ldots, u_n$ or $x_1, \ldots, x_n$.

Let the metric

$$ds^2 = \sum_{i,j=1}^{n} g_{ij}(u)du^i du^j, \qquad u \in D \tag{6.1}$$

be given in some domain D of space $R_u^n, n \ge 2$.

Definition. The metric (6.1) is called *locally Euclidean* if each point of the domain D has a neighbourhood U admitting an isometric embedding into R_x^n with the standard Euclidean metric $ds^2 = \sum_{i=1}^{n} dx^{i^2}$.

In the two-dimensional case $n = 2$ in Chapter 2 we studied the following issues:

(A) Finding the conditions for the coefficients, in the execution of which a metric is locally Euclidean.

(B) Constructing locally isometric embeddings into R_x^n of the neighbourhoods of points from D.

(C) Elucidating the existence of immersions in the large and their smoothness in the large for domains D with the known smoothness of the boundary.

(D) Elucidating if an existing isometric immersion $x : D \to R_x^n$ is an embedding.

The same questions are also naturally to be raised in the general case. The answers to the first two issues are given in [190] for metrics of class C^1. We describe the solution of the problem proposed in that work.

6.1.1   A criterion of local Euclidity in the n-dimensional case

Apparently, if a metric is locally Euclidean, then, after a parallel transport of any vector in a simply connected domain along an arbitrary closed contour, the vector returns to the initial position. In [190] *this property of the parallel transport of vectors is taken as the definition of a C^1-smooth l. E. metric, and it has been proved that in this case there exists a C^2-smooth change of coordinates leading ds^2 to a form with constant coefficients.* Let us lay down this proof.

Let in a domain $D \subset R_u^n$ a C^1-smooth metric (6.1) be given. Parallel translation of the vector T with the coordinates T^i in this metric along any C^1-smooth curve $u = u_i(t)$ is given as the solution of the following system of equations:

$$(T^i)' = -\Gamma_{ks}^i T^s (u^k)',$$

where $'$ denotes differentiation and Γ_{ks}^i are first-kind Christoffel coefficients for the metric considered.

Let $M_0(u^0)$ be some point of the domain D and let a vector T be given at this point. We translate this vector in the parallel way to neighbouring points. By the condition the result of the translation does not depend on the route, n single-valued functions $T^i(u_1, \ldots, u_n)$ – coordinates of the vector T – will emerge in the neighbourhood of the point M_0 in this system of coordinates. For each of these points, we have the relation

$$(\frac{\partial T^i}{\partial u^k} + \Gamma^i_{k\alpha} T^\alpha) du^k = 0.$$

As this equality is true for any set of differentials du^k, it leads to a system of n^2 linear equations with continuous coefficients

$$\frac{\partial T^i}{\partial u^k} = -\Gamma^i_{k\alpha} T^\alpha, \ 1 \leq i, k \leq n. \tag{6.2}$$

In the class of C^1-smooth coefficients (i.e., when the metric is C^2-smooth) this system is solvable under usual consistency conditions of its equations, which are reduced to the satisfaction of the equalities

$$\frac{\partial \Gamma^i_{lj}}{\partial u^k} - \frac{\partial \Gamma^i_{kj}}{\partial u^l} + \Gamma^i_{sk}\Gamma^s_{jl} - \Gamma^i_{sl}\Gamma^s_{jk} = 0,$$

which is equivalent to the equality to zero of all components of the curvature tensor of the metric (6.1). For the case of the system (6.2) with continuous coefficients, [77] showed that the necessary and sufficient condition of the solubility of the system can be represented in the following integral form (in which there is no summation for k and l)

$$\oint_L (\Gamma^i_{jk} du^k + \Gamma^i_{jl} du^l) = \iint_\Omega (\Gamma^i_{\alpha l}\Gamma^s_{jk} - \Gamma^i_{sk}\Gamma^s_{jl}) du^k du^l, \tag{6.3}$$

where L denotes any positively oriented rectifiable Jordan curve situated in the sufficiently small neighbourhood of the point M_0 on the two-dimensional plane with the constants $u^i, i \neq k, \ i \neq l$, and Ω is the domain bounded by this curve (the theory for solving such *systems of equations in total differentials* is also expounded in [73], Chapter VI). Therefore, *equations (6.3) represent the necessary and* *sufficient conditions of local Euclidity of a C^1-smooth metric in the sense of its above accepted definition.*

Such metrics are reduced to the standard form as follows. Consider the system of $n(n+1)$ equations

$$\frac{\partial V_i}{\partial u^k} = \Gamma^j_{ik} V_j,$$
$$\frac{\partial v}{\partial u^k} = V_k \quad (1 \le i, k \le n) \tag{6.4}$$

for $n+1$ functions $v(u^1, \ldots, u^n)$, $V_i(u^1, \ldots u^n)$. This system also belongs to systems of equations in total differentials, and for it the solvability criterion written out in accordance with general theory exactly coincides with the equations (6.3), which are assumed to be performed. Therefore, the system (6.4) is solvable in the class of C^1-smooth functions, and, as the author writes, the component $v(u^1, \ldots, u^n)$ of its solution *unexpectedly* proves to be C^2-smooth. Indeed, $\exists \dfrac{\partial^2 v}{\partial u^i \partial u^k} = \dfrac{\partial V_k}{\partial u^i} = \Gamma^j_{ki} V_j = \Gamma^j_{ik} V_j = \dfrac{\partial^2 v}{\partial u^k \partial u^i}$. Then the system (6.4) proves to be equally matched to the system

$$\frac{\partial^2 v}{\partial u^i \partial u^k} = \Gamma^j_{ki} \frac{\partial v}{\partial u^j}, \tag{6.5}$$

solving which when choosing n independent initial conditions of the form $\partial v(M_0)/\partial u^i = c_i$ (the values $v(M_0)$ in all cases can be considered to equal zero) we obtain n independent C^2-smooth functions $v^1, \ldots, v^n$. The equations of the system (6.5) signify that, as the C^2-smooth transformation $v^i = v^i(u)$ with non-zero Jacobian is passing from coordinates $u = (u^1, \ldots, u^n)$ to coordinates $v = (v^1, \ldots, v^n)$, the Christoffel symbols become equal to zero. This means that the coordinates v^i are affine, hence ds^2 in new coordinates turns to a form with constant coefficients and, after some linear transformation, acquires the standard form. Therefore, the *local reduction of the l. E. metric (6.1) to an Euclidean form is performed through the use of n independent solutions v^i of the system (6.5) as new coordinates.*

In the meantime, the possibility of such a transformation can be substantiated in a slightly shorter way, if we use the following reasoning. As the metric (6.1) is by condition locally Euclidean in the sense of the initial definition and is of smoothness C^1, then by

Theorem 1.1 the existing isometry f between the metric (6.1) and the standard Euclidean metric is of C^2 (and this fact, unknown at the time of writing of [190], explains that there is nothing "unexpected" in the C^2-smoothness of the component $v(u^1, \ldots, u^n)$ of the solution of the system (6.4)). This means that we can calculate the new Christoffel coefficients $\tilde{\Gamma}^k_{ij}$ by the known law of transformation at the change of coordinates and from the condition of their equality to zero to obtain equations for the isometry

$$f = \{f^1, \ldots, f^n\} : \frac{\partial^2 f^k}{\partial u^i \partial u^j} = \Gamma^l_{ij} \frac{\partial f^k}{\partial u^l}, \tag{6.6}$$

which exactly coincides with the system (6.5). But the undoubted benefit of the reasoning performed in [190] is that the *new* definition of the local Euclidity of a metric used in [190] leads to its local Euclidity in the previous sense, and then the C^2-smoothness of transition from the coordinates (u) to the coordinates (v) proves not an unexpected but an obligatory property of the isometry between C^1-smooth isometric metrics.

As in any compact domain D_0 for the isometry $f : D_0 \to R^n$ we have $det(f'(u)) = \sqrt{det(g_{ij})} \geq \min_{\bar{D}_0} \sqrt{det(g_{ij})} > c > 0$, the C^2-smooth solution of the system (6.6) giving an isometric immersion of the metric (6.1) into R^n in a neighbourhood of some point of the domain D, will also be an isometry in each compact subdomain of its existence in D. And as there are locally no obstacles for extending the solution, the extension being a unique one, and the theorem on monodromy is valid, we obtain the following proposition:

Proposition 6.1. *Any domain $D \subset R^n_u$ with an l. E. metric, homeomorphic to an n-dimensional ball, is isometrically immersible into R^n_x. If herewith the metric (6.1) has the smoothness of class $C^{k,\alpha}$, $k \geq 0, 0 \leq \alpha \leq 1, k+\alpha > 0$, in the open domain D, the immersion will be of smoothness class $C^{k+1,\alpha}$; and, in addition, if the domain D has the boundary of smoothness $C^{k+1,\alpha}$, then the smoothness of immersion $\bar{D}$ will be the same.*

The difference of the cases of smoothness of class C^1 and higher from the smoothness $C^{0,\alpha}$, $\alpha > 0$, is that for the former case we are aware of the algorithm of finding the isometry, realized as the solution of an explicitly written system of differential equations (6.6), and in the latter case we only know of its existence. The verbal description of the algorithm, though, is known in this case, too: for manifolds of bounded curvature, without any *a priori* conditions for the character of smoothness of the metric (even without the condition of, say, its continuity), [113] and [114] describe the method of introducing the so-called *distance* coordinates, which in reality prove harmonic coordinates, in which the l. E. metric should prove analytical. A more detailed exposition of these and related problems can be found in [2, 15].

Following V.A. Alexandrov [4] so as not to single out a special class of metrics (6.1) with continuous coefficients, we give the following general definition of metrics' smoothness: a metric (6.1) will be called a *locally Euclidean metric of class* $C^{m,\alpha}$ $(m \geq 1, 0 \leq \alpha \leq 1)$ in the domain D, if each point $u \in D$ will have its neighbourhood U and an isometric embedding $x_U : U \cap D \to R^n_x$ of class $C^{m,\alpha}$. Then the coefficients of the metric (6.1) will in fact be of class $C^{m-1,\alpha}$, including the case $m = 1, \alpha = 0$, in which, as we known, an isometric immersion into R^n does not always has the smoothness C^1 – this is some sharpening of the requirement of metric's smoothness, but instead we achieve the consistency of the conditions of smoothness for all classes of metrics considered. Let us give another (non-constructive) proof of the above Proposition 6.1, also valid for the case of metrics of class $C^{0,\alpha}, 0 \leq \alpha \leq 1$.

Proposition 6.2. *Each locally Euclidean metric (6.1) of smoothness $C^{m,\alpha}(m \geq 1, 0 \leq \alpha \leq 1)$, given in a simply connected domain $D \subset R^n_u$, admits an isometric immersion into R^n_x of class $C^{m,\alpha}$ at $m + \alpha > 0$ and of class $C^{0,1}$ at $m = \alpha = 0$.*

Proof. We fix the point $u_0 \in D$ and such its neighbourhood U_0 that there exists an isometric embedding $x_0 = x_{U_0} : U_0 \cap D \to R^n_x$ of class C^m. For an arbitrary point u of the domain D we construct a finite set of points $u_i \in D$ $(i = 1, 2, \ldots, N)$ and their open neighbourhood $U_i \subset D$ such that 1) $u_N = u$; 2) for each i, there exists an isometric embedding $x_i = x_{U_i} : U_i \cap D \to R^n_x$ of class $C^{m,\alpha}$; 3) for each i, the intersection $U_{i-1} \cap U_i$ is a non-empty connected set.

From the isometricity of the mappings x_{i-1} and x_i, it follows that at each point of the domain of definition of the mapping $x_{i-1} \circ$

$x_i^{-1} : x_i(U_i \cap U_{i-1}) \to R_x^n$ its upper and lower extensions are equal to unity, so its differential is an isometric mapping. Then by the stability theorem of Yu.G. Reshetnyak ([135], p. 20) we obtain that the mapping itself is an isometry too. And since the set $x_i(U_i \cap U_{i-1})$ is open and connected, then a $P_i : R_x^n \to R_x^n$ will be found, whose restriction on the set $x_i(U_i \cap U_{i-1})$ coincides with the mapping $x_{i-1} \circ x_i^{-1}$. In other words, we found such an isometry P_i, that for all $u \in U_{i-1} \cap U_i$ the equality $P_i[x_i(u)] = x_{i-1}(u)$ is valid. Finally, let us construct a mapping $x : D \to R_x^n$, setting its value at the point u by the equality

$$x(u) = (P_1 \circ P_2 \circ \cdots \circ P_{N-1} \circ P_N \circ x_N)(u).$$

The mapping x is defined correctly in the sense that its value at the point u does not depend on the choice of auxiliary points u_i and their neighbourhoods U_i, because the domain D is simply connected and, therefore, the theorem of monodromy takes place. On the other hand, x by definition is an isometric immersion of the metric (6.1), and by the condition of the theorem is C^m-smooth. And if $ds^2 \in C$ and, nevertheless, $x_U \in C^1$, then, certainly, the immersion in the large will be of class C^1.

Remark 6.1. As the concepts of the upper and lower extensions are also correctly defined for the class of Lipschitz mappings (see the formula (6.7) below), then instead of the assumption $x_U \in C^1$ it is sufficient to have the condition $x_U \in C^{0,1}$, therefore the adduced proof is performed for any continuous metric too, but in this case the immersion can be of class $C^{0,1}$.•

6.1.2   *Embeddings of n-dimensional locally Euclidean metrics into R^n*

The proposition below on the isometricity of a simply connected domain with a complete l. E. metric to space R^n with the standard Euclidean metric is a particular case of the Killing–Hopf theorem (see Corollary 2.4.10 in [192]), but at the same time formally this proposition makes more exact the respective part of the Killing–Hopf theorem, as, first, it also deals with l. E. metrics given simply by a continuous metric form; second, it gives information on the class of smoothness of the performed embedding, thus solving completely the problem of embeddability into R_x^n of a complete locally Euclidean metric given in the entire space R_u^n. In classes of smoothness C^2 and higher, with the assumption that in R_u^n there exists a point p, through which n linearly independent "straight lines" pass (i.e., complete geodesics, each arc of which is a shortest path), the isometricity of R_u^n with a complete l. E. metric to usual Euclidean space R_x^n also follows from the results of [177, 178] and [75].

Theorem 6.1. *Let a locally Euclidean metric (6.1) of class C^m, $m \geq 1$, be given in all space R_u^n, $n \geq 2$. Then*
   (a) if the metric (6.1) is complete, then it admits an isometric embedding $x : R_u^n \to R_x^n$ of class C^m; herewith, $x(R_u^n) = R_x^n$;
   (b) if for some $l \geq 1$ the metric (6.1) admits an isometric embedding into R_x^n of class C^l, where $x(R_u^n) = R_x^n$, then it is complete. (See [4].)

For the proof of the theorem, we will need a definition and a variant of the theorem of global inverse function, which use the notation: B and b are Banach spaces; D is an open and connected subset of space B; $x : B \to b$ is a continuous mapping.

   Following R. Plastok [123], we will say that the mapping x satisfies the condition (L), if for any points x_1, x_2 of space b and any path $P : [0, T) \to D$, $T < \infty$, satisfying for all $t \in [0, T)$ the relation

$$x(P(t)) = L(t),$$

where $L(t) = [(T - t)x_1 + tx_2]/T$, a sequence of points t_j will be found, which at $j \to \infty$ tends to T on the left-hand side, such that $\lim_{j \to \infty} P(t_j)$ exists and lies in D.

Proposition 6.3. *Let $x : D \subset B \to b$ be a local homeomorphism. Then the condition (L) is necessary and sufficient for x to be a homeomorphism of D on b. (See [123], the theorem on the global inverse function.)*

Proof of Theorem 6.1 [4]. (a) According to Proposition 6.2, for the metric (6.1) there exists its isometric immersion $x : R_u^n \to R_x^n$ of class C^1. Let T, L and P have the same meaning as in the above definition of the condition (L). An arbitrary sequence of numbers, tending at $j \to \infty$ to T from the left, will be taken as t_j. The distance between the points $P(t_i)$ and $P(t_j)$ in the metric (6.1) does not exceed the length of the segment of the curve $P(t)$ between them, measured in the metric (6.1). The metric, owing to the local isometricity of the mapping x, is equal to the length of the segment of the curve $x(P(t)) = L(t)$, enclosed between the points $x(P(t_i))$ and $x(P(t_j))$ (this time, the length is measured in the Euclidean metric in R_x^n). But $L(t)$ is a straight line, so the length of the latter curve is equal to the Euclidean distance between the points $x(P(t_i)) = L(t_i)$ and $x(P(t_j)) = L(t_j)$. Finally, as $L(t_j) \to L(T) = x_2$ when $j \to \infty$, the sequence of points $P(t_j)$ is fundamental in the metric (6.1). On the strength of the completeness of the metric (6.1), the sequence $P(t_j)$ converges in this metric to some point of space R_u^n. But then this sequence will also converge to the same point in the Euclidean metric in R_u^n. Therefore, the immersion x possesses the property of (L), and so, by the Plastok theorem 6.3 it is a homeomorphism "on" or, in other words, such an embedding that $x(R_u^n) = R_x^n$. Proposition (a) of Theorem 6.1 is proved. The proof of Proposition (b) is evident.

In reality, the same theorem can be easily proved by simpler reasoning, and in a somewhat more general formulation too, by assuming the complete metric given only in some ball $D : |u| < R$, $0 < R \leq \infty$[48].

Proof. First we need again, in fact, to prove only part (a). As for the metric (6.1) there is its immersion $x : D \to R_x^n$ of smoothness at least C^1, then the segments of the straight lines from the image in the pre-image are assigned C^1-smooth geodesic lines going out of each point in a certain direction and uniquely determined by that direction. Owing to the completeness of the metric, each geodesic extends unboundedly, therefore, a complete straight line corresponds to it in R_x^n. We fix in D some point M_0 and release from it geodesic rays in each direction. In R_x^n, they will be assigned rectilinear rays, emanating from $x(M_0) \in R_x^n$, and in totality covering the entire space. The geodesic rays emanating from $M_0 \in D$ also cover the entire ball D. Indeed, let an arbitrary point $M_u \in D$ be given. By the

---

[48] Obviously, the conditions of setting the metric in a ball or in the whole space are in reality reducible one to the other without loss of completeness of the metric.

immersion $x : D \to R_x^n$, in R_x^n it is assigned some point M_x. We cover the segment $[x(M_0), M_x]$ in R_x^n by a finite system of neighbourhoods U_i, starting from the neighbourhood $U_0 \ni x(M_0)$, in each of them for the mapping x there exists an inverse bijective mapping uniquely defined by the initial pair $M_0 \leftrightarrow x(M_0)$. The union of the pre-images of the segments of the straight line $x(M_0)M_x$, contained in U_i, gives in D an arc of the geodesic ray coming from M_0 into M_u. After this is established, it becomes evident that it could not be that two points M_1 and M_2 from D have a coinciding image in R_x^n. This means that the surjection $x : D \to R_x^n$ is in reality a bijection, so the immersion $x(D)$ in reality yields the embedding of D into R_x^n with $x(D) = R_x^n$.

Note that from these proofs it follows that *each* isometric immersion x into R_x^n of a complete locally Euclidean metric, given in an arbitrary simply connected domain D in R_u^n, is in reality an embedding such that $x(D) = R_x^n$.

If we omit the condition of completeness of the metric, then the natural problem would be to find any necessary or sufficient conditions of embeddability of the metric into R_x^n. Below we will obtain the sufficient condition of embeddability into R_x^n of a locally Euclidean metric (6.1) in terms of the increase of its coefficients at infinity. For this we will need one more variant of the global inverse function, for the formulation of which we will introduce the following notation.

If B and b are Banach spaces, and the mapping $x : B \to b$ is a local homeomorphism, then the lower dilatation [93] of the mapping x at the point $X \in B$ will be designated as

$$D_X^- x = \liminf_{Y \to X} \frac{|x(Y) - x(X)|}{|Y - X|}, \qquad (6.7)$$

where $\liminf$ and $|\cdot|$ are the symbols of the lower limit and the norms, respectively. Note that if x is differentiable at the point X, and the differential $x'(X)$ is inversible, then, evidently, $D_X^- x = \|x'(X)^{-1}\|^{-1}$, where $\|\cdot\|$ is the operator norm of the linear mapping. Besides, for non-negative real numbers t, we define the non-increasing function M, setting it by the equality

$$M(t) = \inf_{\substack{|X| \leq t \\ X \in B}} D_X^- x.$$

Proposition 6.4. (On the global inverse function; [30, 68, 93, 126] and others.) *Let B and b be Banach spaces, and the mapping $x : B \to b$ be a local homeomorphism; herewith,*

$$\int\limits_{0}^{+\infty} M(t)\, dt = +\infty.$$

Then x maps B bijectively onto all space b.

Lemma 6.1. *Let $A = (a_{ij})$ be a nongenerated real $n \times n$ matrix. Then*

$$\|A^{-1}\|^2 \leq \left(n \sum_{\substack{i=1}}^{n} \prod_{\substack{j=1 \\ j\neq i}}^{n} g_{jj} \right) \Big/ \det(g_{kl}),$$

where $g_{kl} = \sum_{s=1}^{n} a_{ks} a_{ls}$, and $\|\cdot\|$ is the operator norm of the matrix constructed by the Euclidean norm in $\mathbb{R}^n$.

Proof. Let A_{ij} be the cofactor of the element a_{ij} of the matrix A and $\Delta = \det A$. Then

$$\|A^{-1}\|^2 = \|A_{ik}/\Delta\|^2 \leq \sum_{i,k=1}^{n} A_{ik}^2/\Delta^2 \leq$$

$$\leq \sum_{i,k=1}^{n} \left[\prod_{\substack{j=1 \\ j\neq i}}^{n} \sum_{l=1}^{n} a_{jl}^2 \right] \Big/ \Delta^2 = \left(n \sum_{i=1}^{n} \prod_{\substack{j=1 \\ j\neq i}}^{n} g_{jj} \right) \Big/ \det(g_{kl}).$$

The first of the written inequalities is valid because the squared norm of the matrix does not exceed the sum of the squares of its elements; the second, due to the known Adamar inequality for the determinants. The lemma is proved.

Theorem 6.2. *Let in R_u^n $(n \geq 2)$ a locally Euclidean metric (6.1) of class C^m $(m \geq 1)$ be given. Let, besides, there exist a non-increasing continuous function $M : [0, +\infty) \to (0, +\infty)$ such that*

$$\int\limits_{0}^{+\infty} M(t)\, dt = +\infty \tag{6.8}$$

and

$$\left(n \sum_{\substack{i=1}} \prod_{\substack{j=1 \\ j\neq i}}^{n} g_{jj}(u) \right) \Big/ \det(g_{kl}(u)) \leq [M(|u|)]^{-2} \qquad (6.9)$$

for all $u \in \mathbb{R}^n_u$. Then the metric (6.1) admits an isometric embedding $x : R^n_u \to R^n_x$ of class C^m; herewith, $x(R^n_u) = R^n_x$.

Proof. As we know, for the metric (6.1) there exists an isometric immersion $x : R^n_u \to R^n_x$ of class C^m. By Lemma 6.1

$$D_u^- x = \|x'(u)^{-1}\|^{-1} \geq M(|u|).$$

Then by Proposition 6.4 we have that x is a homeomorphism for all R^n_x. The theorem is proved.

Note that in reality we have proved that under conditions of the theorem *any* isometric immersion of the metric (6.1) into R^n_x is an embedding. As in this case the entire space is an image, the metric is complete, so the conditions of the theorem can be also treated as the sufficient conditions of the completeness of the given metric.

Note also that the condition (6.8) for the diminution of the function $M(u)$ in the theorem is unimprovable. To be more exact, for each $n \geq 2$ and for each non-increasing continuous function $m : [0, +\infty) \to (0, +\infty)$ such that

$$\int_0^{+\infty} m(t)\, dt < +\infty,$$

there exist a locally Euclidean metric (6.1) in R^n_u and the numbers $c > 0$ and $t_0 > 0$ such that for all $t \geq t_0$ the following inequality is satisfied:

$$c\, m(t) \leq M(t) \equiv \inf_{|u| \leq t} \frac{(\det(g_{kl}(u)))^{1/2}}{(n \sum_{\substack{i=1}} \prod_{\substack{j=1 \\ j \neq i}}^{n} g_{jj}(u))^{1/2}}.$$

However, this metric does not admit an isometric embedding into $\mathbb{R}^n_x$.

Indeed, John ([93], p. 92) constructs for any such function m a continuously differentiable mapping $x = (x^1, x^2) : \mathbb{R}^2_u \to \mathbb{R}^2_x$:

(a) x is surjective but not injective;

(b) for all $u \in \mathbb{R}^2_u$ the rows of the matrix $x'(u)$ are pairwise orthogonal and $\det x'(u) > 0$;

(c) for each $t \geq 0$ we have $\inf\limits_{|u| \leq t} D^-_u x = m(t)$.

Let us set the mapping $X = (X^1, X^2, \ldots, X^n) : \mathbb{R}^n_u \to \mathbb{R}^n_X$ by the formulae

$$
\begin{aligned}
X^1(u^1, u^2, \ldots, u^n) &= x^1(u^1, u^2), \\
X^2(u^1, u^2, \ldots, u^n) &= x^2(u^1, u^2), \\
X^3(u^1, u^2, \ldots, u^n) &= u^3, \\
&\cdots \\
X^n(u^1, u^2, \ldots, u^n) &= u^n.
\end{aligned}
$$

It is easy to see that the mapping X possesses the following properties:

(A) X is surjective but not injective;

(B) for all $u \in \mathbb{R}^n_u$ the rows of the matrix $X'(u)$ are pairwise orthogonal and $\det X'(u) > 0$;

(C) if the number t_0 is such that $m(t_0) < 1$, then for all $t \geq t_0$ we have

$$
M(t) \geq c \inf\limits_{|u| \leq t} D^-_u X \geq c\, m(t),
$$

where c is some constant.

Only the property (C) requires to be proved. The left inequality in it occurs because, if we assume $A = X'(u)$ and make use of the notation of Lemma 6.1, we will have

$$
\left(n \sum_{\substack{i=1 \\ j=1 \\ j \neq i}}^{n} \prod g_{jj}(u) \right) \Big/ \det(g_{kl}(u)) = \sum_{i,k=1}^{n} \left[\prod_{\substack{j=1 \\ j \neq i}}^{n} \sum_{l=1}^{n} a^2_{jl} \right] \Big/ \Delta^2 =
$$

$$
= \sum_{i,k=1}^{n} A^2_{ik}/\Delta^2 \leq c^2 \, \|(A_{ik}/\Delta)\|^2 = c^2 \, \|A^{-1}\|^2 = c^2 / (D^-_u X)^2.
$$

The second equality is written here because due to the property (B) the rows of the matrix $A = X'(u)$ are pairwise orthogonal, and for such matrices the Adamar inequality becomes an equality. The only inequality and the constant c here simply express the equivalency of two matrix norms. To prove the right-hand-side inequality of the property (C), it is sufficient to solve some elementary inequality, which we leave to the reader.

We transfer the Euclidean metric $(dX^1)^2 + (dX^2)^2 + \cdots + (dX^n)^2$ of the space $\mathbb{R}^n_X$ to the space R^n_u of the variables $u^1, u^2, \cdots, u^n$ using the mapping X. The obtained metric is, apparently, locally Euclidean and satisfies the inequality $c\, m(t) \le M(t)$ for $t \ge t_0$. Besides, one of its immersions (namely, X) is not an embedding. But from the proof of Proposition 6.2 it follows that any other of its immersions has the form $P \circ X$, where $P : R^n_X \to R^n_X$ is some isometry. Therefore, no immersion of the constructed metric is an embedding.

Note that the conditions of Theorem 6.2 are not invariant relative to coordinate substitution. Indeed, if $u = u(v^1, v^2, \cdots, v^n) : R^n_v \to R^n_u$ is a diffeomorphism, then using it we can transfer the metric (6.1) to the space of variables $v^1, v^2, \ldots, v^n$ (we designate the transferred metric as $d\sigma^2$). However, it is easy to see that the diffeomorphism u can always be chosen such that for the metric $d\sigma^2$ the conditions of Theorem 6.2 are not satisfied.

From the geometrical point of view, this non-invariance of the conditions of Theorem 6.2 reduces its value and makes topical the consideration of the following circle of problems:

Open question 6.1. 1) to find conditions for the coefficients of the metric (6.1), invariant relative to the change of coordinates and guaranteeing the embeddability of the metric into R^n_x;

2) for a given metric (6.1), to find the best (from the point of view of Theorem 6.2) coordinate system, i.e., such a coordinate system, in which for the function M, satisfying the condition (6.8), the integral

$$\int_0^{+\infty} M(t)\, dt \tag{6.10}$$

takes the largest value;

3) to study whether Theorem 6.2 remains valid if, instead of the condition (6.8), we require that the exact upper boundary of the

integrals (6.10), considered for all coordinate systems and all functions M satisfying the condition (6.8), be equal to infinity.$\bullet$

Note that in the two-dimensional case the most complete results were obtained for embeddings of two-dimensional metrics given in isothermal coordinates. If we succeeded in applying Theorem 6.2 to such a metric, then for any its embedding x we would have had $x(R_u^2) = R_x^2$, but then due to Theorem 6.1 the metric would have been complete. But the proposition of Corollary 4.2 shows that the form of complete metrics on the plane in isothermal coordinates is exhausted by the trivial generalization of the standard Euclidean metric, so Theorem 6.2 is ill-suited for metrics defined in isothermal coordinates and works in another class of metrics. This remark also pertains to the multidimensional case, as the following proposition is also valid there:

Proposition 6.5. *Any locally Euclidean metric of class C^m ($m \geq 1$), given in isothermal coordinates in R_u^n ($n \geq 3$), has constant coefficients.*

Proof. It is evident that any isometric immersion x of this metric into R_x^n is simultaneously a conformal mapping of R_u^n into R_x^n. Then by the Liouville theorem [135] the mapping x is Möbius one, and, thus, linear. The proposition is proved.

Further, our aim is to prove the below Theorem 6.3 on the radius of the injectivity of immersion of a locally Euclidean metric. This proof is based on the following variant of the global inversion function theorem proved by John [93] and presented here in a convenient formulation.

Proposition 6.6. (Theorem IIA, [93].) *Let B and b be Banach spaces, x be a locally homeomorphic mapping of the ball of space B with the centre u of radius R into space b. Assume for $0 \leq t < R$*

$$M(t) = \inf_{|u-v| \leq t} D_v^- x.$$

Then in the ball ω of space b with the centre $x(u)$ of radius

$$r = \int_0^R M(t)\, dt$$

the mapping x^{-1} is defined, which is inverse to x, i.e., such that
    (a) $\omega \subset x(B)$;
    (b) x^{-1} is continuous in ω;
    (c) $x(x^{-1}(v)) = v$ for all $v \in \omega$.

Now we can prove the following theorem on the embedding of l. E. metrics.

Theorem 6.3. *Let in $\mathbb{R}^n_u$ $(n \geq 2)$ a locally Euclidean metric (6.1) of the class of smoothness C^m $(m \geq 1)$ be given. For an arbitrary point $u \in R^n_u$ we construct the functions*

$$N(t) = \sup_{|v-u|\leq t} V(v) \qquad and \qquad M(t) = \inf_{|v-u|\leq t} W(v),$$

where

$$V(v) = \left[\prod_{i=1}^{n} g_{ii}(v)\right]^{1/2},$$

$$W(v) = [\det(g_{kl}(v))]^{1/2}\left[n \sum_{\substack{i=1 \\ }}^{n} \prod_{\substack{j=1 \\ j\neq i}}^{n} g_{jj}(v)\right]^{-1/2}.$$

Then for any positive r satisfying the inequality

$$rN(r) < \int_{0}^{+\infty} M(t)\, dt, \tag{6.11}$$

the Euclidean ball with the centre in $u \in R^n_u$ and of radius r, equipped with the metric (6.1), admits an isometric embedding into $\mathbb{R}^n_x$ of class C^m.

Proof. We are aware that for the metric (6.1) there exists its isometric immersion $x : \mathbb{R}^n_u \to \mathbb{R}^n_x$ of class C^m. Due to Lemma 6.1

$$D_v^- x = \|x'(v)^{-1}\|^{-1} \geq W(v) \geq \inf_{|w-u|\leq|v-u|} W(v) = M(|v-u|).$$

Then by Proposition 6.6, in a Euclidean ball ω of the space $\mathbb{R}_x^n$ with the centre $x(u)$ and radius

$$\rho = \int_0^{+\infty} M(t)\, dt$$

a mapping inverse to x is defined. On the other hand, for any positive r satisfying the inequality (6.11), and any $v \in \mathbb{R}_u^n$ such that $|v-u| < r$, we have

$$|x(v) - x(u)| \leq |v - u| \sup_{|w-u|\leq r} \|x'(w)\| \leq$$

$$\leq r \sup_{|w-u|\leq r} \left\{ \prod_{i=1}^n \sum_{j=1}^n \left[\frac{\partial x^i}{\partial u^j}(w) \right]^2 \right\}^{1/2} = r \sup_{|w-u|\leq r} \left[\prod_{i=1}^n g_{ii}(w) \right]^{1/2} =$$

$$= rN(r) < \int_0^{+\infty} M(t)\, dt = \rho$$

(the second inequality here is a corollary of the Adamar inequality). This means that the image of the ball $|v - u| < r$ at the mapping of x is contained in the ball ω. Therefore, the restriction of the mapping of x onto the ball $|v - u| < r$ is injective. The theorem is proved.

6.1.3   *Surfaces with locally Euclidean metrics in $E^n, n > 3$*

Already in the case of E^4 we can see that the local structure of surfaces with an l. E. metric can differ significantly from their familiar structure in E^3, where the main characteristic of such surfaces was the existence of rectilinear generatrices on them. An example of the complete Möbius band (5.15) located in a three-dimensional ellipsoid shows that the generatrices can be also absent on noncompact surfaces. Therefore, to single out "customary" surfaces with an l. E. metric in multidimensional spaces, it is necessary to introduce some additional conditions.

6.1.3.1   *Ruled surfaces with locally Euclidean metrics.*   The structure of a ruled surface with an l. E. metric can be considered as a natural generalization of developable surfaces in the three-dimensional

space, the more so that under some codimensionality conditions complete surfaces with an l. E. metric inevitably prove to have a ruled structure. For formulation of respective theorems, recall some definitions and terms (here our exposition predominantly follows [179, 182].

The *relative* (or *external*[49]) *nullity index* $\nu(P)$ of a given l-dimensional surface $F^l \subset E^n, 2 \le l < n$, at some its point P is a maximum dimension ν of the subspace T_P° of those vectors X from the tangent space $T_P(F^l)$ of the surface, for which the vector-valued second form $A(X, Y)$ of the surface with values in normal space is equal to zero for all vectors $Y \in T_P(F^l)$. In other words, if the second form is represented as a set of $n - l$ bilinear forms $A_{ij}^r X^i Y^j$ for each vector of the normal from an orthogonal basis in the normal space of the surface, then $A_{ij}^r X^i = 0$ for all vectors $X = \{X^i\} \in T_P^\circ$ and all $r, 1 \le r \le n - l$. In particular, if we take $Y = X \in T_P^\circ$, then $A(X, X) = 0$, i.e., all vectors of the subspace T_P° are asymptotic (the inverse is, generally speaking, false). This subspace T_P° is called the *space of relative* or *external nullity space* of the surface at the point P.

We give the following:

Definition 1. A surface F^l in E^n is called *developable*, if its nullity index at each point is equal to $l - 1$.

In particular, at each point of such a surface there exist at least $(l - 1)$ independent asymptotic directions.

Definition 2. The surface $F^l \subset E^n$ is called k-*ruled*, if through each point $P \in F^l$ passes the *unique* k-dimensional plane domain $L \subset F^l$, boundary points of which belong to the boundary of the surface (if any).

By analogy with ruled surfaces in E^3, plane domains L are called k-dimensional plane generatrices (rectilinear generators). We will assume that the indicated structure of the surface is preserved for an arbitrarily small domain of the surface; we will call such surfaces *strictly k-ruled*. At the beginning of Subsection 3.1.7 we give an example of a surface, which is 1-ruled in the sense of Definition 2, but not strictly ruled, because at a sufficiently strong decrease of the

---

[49] There is also the notion of the *internal* nullity index, which in our case is equal to l.

neighbourhood of any point M_0 from the triangle considered there the property of uniqueness of the generatrix passing through the point M_0 vanishes. Apparently, the definition of strict k-ruledness is equivalent to the requirement of the absence of points from the $(k+1)$-plane neighbourhood on the surface.

It proves that $(l-1)$-ruled surfaces and developable surfaces are related as follows:

Theorem 6.4. *For a C^2-smooth surface $F^l \subset R^n$ to be developable, it is necessary that it be strictly $(l-1)$-ruled with the stationary tangent plane along each generatrix; the same property of a surface is sufficient for it to be everywhere developable with the possible exception of some set of points of l-dimensional measure zero (which is defined by the condition that the nullity index at its points is equal to l).*

Proof of the theorem[50] with respect to the necessity can be found, for instance, in [75], Lemma 3.1; [26], p. 225; in [179] and [182]. That exceptional manifold on the surface, in which the nullity index can be equal not to $(l-1)$, but to l, corresponds to points of flattening, and it should be absent on a developable surface by its definition. If points with nullity indices $l-1$ and l occur on a surface, the surface is called a developable surface with planar points, and for such surfaces [179] gives suitable descriptions of the structure of their generatrices; [75] in Lemma 3.1 also considers cases of non-constant nullity index.

By analogy with surfaces in the three-dimensional space, an important point in the proof of the necessity is not only the establishment of the strictly $(l-1)$-ruled geometry of a developable surface with the stationary tangent plane along the generatrices, but also the possibility of introducing on a developable surface (see [75]) an asymptotic parametrization

$$r(u^1,\ldots,u^l) = \rho(u^l) + \sum_{i=1}^{l-1} u^i a_i(u^l), \qquad (6.12)$$

with respect to the smoothness of which the same propositions are valid as in the case of the three-dimensional space. Namely, for a

---

[50] The theorem is also valid for surfaces with any constant nullity index $\nu > 0$ with respective change in the dimension of plane generatrices, see [75].

$C^k(k \geq 2)$-smooth surface in this parametrization the directrix $\rho(u^l)$ is a C^k-smooth curve, and the directing vectors $a_i(u^l)$ have a C^{k-1} smoothness.

As these conditions are necessary, in the proof of sufficiency it is natural to assume that the considered $(l-1)$-ruled surface is of $C^k, k \geq 2$, class of smoothness with the respective necessary conditions of smoothness in asymptotic parametrization. Let us find in these assumptions the conditions of stationarity of the tangent planes along the generatrices. The tangent planes of the surface at the points $P_0(0,\ldots,0,u_0^l)$ and $P(u^1,\ldots,u^{l-1},u_0^l)$ are linear spans of the tangent vectors in corresponding points. We have

$$\frac{\partial r}{\partial u^i}(P_0) = a_i(P_0) = a_i(P) = \frac{\partial r}{\partial u^i}(P),$$

$$\frac{\partial r}{\partial u^l}(P_0) = \frac{\partial \rho}{\partial u^l}(P_0),$$

$$\frac{\partial r}{\partial u^l}(P) = \dot{\rho}(P) + u^i \dot{a}_i(P),$$

where the dot over the letters implies the derivative with respect to u^l. For the tangent planes $T_{P_0}(F^l)$ and $T_P(F^l)$ to coincide, it is necessary and sufficient that the vectors $\dot{a}_i(P) = \dot{a}_i(P_0)$ belong to the linear span of the vectors $a_1,\ldots,a_{l-1}$ and $\frac{\partial \rho}{\partial u^l}(P_0)$, i.e., that we had the representation of the form

$$\frac{da_i}{du^l}(P_0) = \beta_i(u^l)\frac{d\rho}{du^l}(P) + \beta_i^j(u^l)a_j(u^l), 1 \leq i,j \leq l-1. \qquad (6.13)$$

From equation (6.13) it follows that the vectors $\dfrac{\partial^2 r}{\partial u^i \partial u^l}$ remain on the tangent plane, hence the coefficients of the second form $A_{il}^r(P) = 0, 1 \leq i \leq l-1, 1 \leq r \leq n-l$; what is more, $\dfrac{\partial^2 r}{\partial u^i \partial u^j} = 0, 1 \leq i,j \leq l-1$, and all together this is what implies that at general-position points, at which $A_{ll}^r \neq 0$ at least for some r, the nullity index of the surface is equal to $l-1$.

We should note here that, as the second derivatives of the vectors $a_i(u^l)$ may fail to exist (if the initial smoothness of the surface was of class C^2), then the coefficients $A_{ij}^r, 1 \leq i,j \leq l$, of the second form in asymptotic coordinates should be considered to be obtained from its coefficients $\tilde{A}^r_{ij}$ in some coordinates $(\tilde{u}^1, \ldots, \tilde{u}^l)$, in which $F^l \in C^2$, by corresponding transformation of the quadratic form at the C^1-smooth transition to asymptotic coordinates $(u^1, \ldots, u^l)$; at such a transformation the coefficients A^r_{ij}, which correspond to the existing second derivatives, will coincide with those coefficients, which are obtained by coordinate-substitution transformations from $\tilde{A}^r_{ij}$, and the coefficients of the form A_{ll} will be calculated without using the second derivatives with respect to u^l.

Following Theorem 6.4, we have the following description of surfaces with an l. E. metric:

Corollary 6.1. *For a C^2-smooth surface $F^l \subset R^n$ to be developable, it is necessary that it be strictly $(l-1)$-ruled and have an l. E. metric; the same property of a surface is sufficient for it to be everywhere developable with the possible exclusion of some set of points of l-dimensional measure zero.*

Indeed, if a surface is developable, then by the theorem it is strictly $(l-1)$-ruled with the stationary tangent planes along the generatrices. Then from the representation (6.12) we have that for second-form coefficients all $A^r_{ij} = 0, 1 \le i, j \le l - 1$, and also, as we have just shown, all $A^r_{il} = 0$. Therefore, all components of the curvature tensor R_{ijkm} are equal to zero, including also the components of the form $R_{iljl} = \Sigma^{n-l}_{r=1}(A^r_{ij}A^r_{ll} - A^r_{il}A_{lj}) = 0$, therefore, the metric of the surface is plane. Conversely, if a surface is strictly $(l - 1)$-ruled and has an l. E. metric, then from the equalities $R_{ilil} = -(A^r_{il})^2 = 0$ we have that $A^r_{il} = 0$, but this equality, as we established above, is sufficient for the tangent plane to be stationary; therefore, by the theorem, the surface is developable.

Thus, we gave such a definition of developable surfaces in a multidimensional space, which makes their structure similar to the situation in the three-dimensional space.

6.1.3.2 Description of surfaces with locally Euclidean metrics via a point codimension. Recall that the *first normal space* N_1 for a surface $F^l \subset R^n$ at the point P is defined as the linear span of the image of the vector-valued second fundamental form A of the surface at the point P : $N_1(P) = span(ImA(P))$. The dimension of the first normal space $N_1(P)$ is called the *point codimension* of the surface F^l at the point P with the notation $codim_P F^l$. The smaller the point codimension is, the closer the surface to the plane is. For an l-dimensional plane, the point codimension is equal to zero, and developable surfaces have point codimension 1, which is confirmed by the following theorem [179, 182].

Theorem 6.5. *For a C^2-smooth surface $F^l \subset R^n$ to be developable, it is necessary and sufficient that its metric be flat, and its pointwise codimension be everywhere equal to 1.*

The proof is obtained by establishing the equality $codim_P(F^l) + \nu(P) = l$ in the conditions of the theorem. Unlike the previous theorem, here there are no exceptional points, where the nullity index of the surface could be equal to zero.

The example of a Clifford torus with $codim(F^2) = 2$ shows that the condition of the theorem $codim(F^l) = 1$ is essential for it to be valid. At the same time, this theorem gives as a corollary the following description of surfaces with an l. E. metric:

Corollary 6.2. *Any C^2-smooth hypersurface $F^{n-1} \subset R^n$ without points of flattening and with an l. E. metric is developable.*

If we assume the presence of only zero-measure set points of flattening, the proposition of the corollary can be made more exact as follows:

Corollary 6.3. *Any C^2-smooth hypersurface $F^{n-1} \subset R^n$ with an l. E. metric and with no more than the $n-1$-dimensional zero-measure set of possible points of flattening is a strictly $(n-2)$-ruled surface.*

In [182], we can also find other descriptions of developable surfaces, including possible occurrences of planar points.

Remark 6.2. As in the case of two-dimensional surfaces in E^3, a decrease of smoothness can totally change the structure of even a hypersurface. Namely, in [29] it is pointed out that any hypersurface admits bendings in the class of hypersurfaces of smoothness $C^{1,\alpha}$, $\alpha < \dfrac{1}{1 + n^2(n+1)}$, such that, on the surface, relations are lost between extrinsic and intrinsic geometries, in particular, a hyperplane can be bent such that there will be no plane generatrices on it.•

6.1.3.3 Classification of developable surfaces. We know that in a three-dimensional space analytical developable surfaces are exhausted by three types: cylinders, cones and torses. It turns out that in a multidimensional case for developable surfaces there is also a classification, which makes it possible, at least in the analytical class, to distribute all surfaces by several types. This was made in [179]. For brevity, surfaces F^l with the nullity index $l-1$ will be called *surfaces of rank 1* (there is also a general definition of surfaces of rank $l-k$ as surfaces with nullity index k). First, we recall the general definitions.

Definition 3. A surface $F^{l+k} \subset R^n$ is called a *k-cylinder or k-cylindrical surface over the surface* $F^l \subset R^n$, if it consists of all k-dimensional planes E^k, drawn through each of the points of the surface F^l in a parallel manner to a given plane Π^k of dimension k. For the surface F^{l+k} thus obtained, the plane Π_k is called the *direction of cylindricity;* the surface F^l, the *directrix,* and the planes E^k are called *k-dimensional generatrices.*

Definition 4. The surface $F^{l+1} \subset R^n$ is called a *1-cone or a 1-conical surface over the surface* $F^l \subset R^n$ *and with the vertex at the point O,* if it consists of all straight lines drawn through each of the points of the surface F^l and the vertex O. For the surface F^{k+l} thus obtained, the surface F^l is called the *directrix,* and the drawn lines the *generatrices.*

It is important to note that a cylinder over a cylinder is also a cylinder; a 1-cone over a 1-cone, it turns out, can be considered as a 1-cylinder over a 1-cone. Indeed, let the surface F^{l+1} be a 1-cone over some surface F^l and with the vertex at the point O. Let the surface F^{l+2} be constructed as a 1-cone over F^{l+1} with the vertex at some point $A \neq O$ (at coinciding A and O the surface F^{l+2} will have the same points as F^{l+1}). Let $C \neq O$ be some point of the cone F^{l+1}. The segment OC consists of points of the surface F^{l+1}, so the 2-dimensional triangle $OAC \subset F^{l+2}$. This means that if the points-vertices O and A are not included into the number of points of the surface F^{l+2}, then it will locally be a 1-cylinder over F^{l+1} with rectilinear segments of the generatrices parallel to the straight line OA.

As the suggested classification bears in mind only the local structure of a surface, this observation justifies the introduction of the following definition.

Definition 5. A surface is called a *k-cylindrocone*, if it is either a *k*-cylinder or a $(k-1)$-cylinder over a 1-cone. In particular, a 1-cylindrocone implies that the surface is either a 1-cylinder or a 1-cone.

Besides cylinders, cones and cylindrocones, the classification of developable surfaces will also involve multidimensional *torses*.

Definition 6. The surface $F^l \subset R^n$ is called an *l-torse*, $l \geq 2$, if it is formed by osculating $(l-1)$-planes[51] of some C^{l+1}-smooth essentially $(l+1)$-spatial curve called the *base* curve of a torse. For a 1-torse, by definition we will take an arbitrary C^2-smooth curve with the curvature unequal to zero.

The essentially *k*-spatiality of a C^k-smooth curve means that the curve lies in no plane of dimension $k-1$, for which it is sufficient that the first k derivatives of its position vector compose a linearly independent system of vectors; the first $(k-1)$ vectors of this system form a $(k-1)$-dimensional tangent plane of the curve.

Writing down the equation for the torse as the equation of a ruled surface and using the Frenet formulae for the base curve, by direct calculations it can be made sure that the torse is a surface of rank 1, or else it is a developable surface. A cylindrocone over a torse is also a surface of rank 1.

After introduction of the above classes of surfaces, Ushakov [179] proves the following theorem on the structure of developable surfaces.

Theorem 6.6. *Any developable surface $F^l \subset R^n$ of smoothness C^{l+1} is either an l-torse or a k-cylindrocone over some $(l-k)$-torse.*

The proof of the theorem is known only from the text of the dissertation [179] and for this reason we will present here its main stages in,

---

[51] At $l = 2$, the tangent straight line of the curve is its osculating 1-plane.

hopefully, sufficient detail for understanding the idea and technique of the proof.

We begin with the consideration of a developable C^3-smooth two-dimensional surface F^2 in R^n. It is a 1-ruled surface. Let $P \in F^2$ be a point of general position. Choosing the C^3-smooth curve $\gamma : \mathbf{r} = \boldsymbol{\rho}(v^2)$ – where v^2 is a natural parameter – as a directrix orthogonal to the generatrices, we will get the representation of the position vector r of the surface F^2 in the form

$$\mathbf{r}(v^1, v^2) = \boldsymbol{\rho}(v^2) + v^1 \mathbf{e}_1(v^2),$$

where $\mathbf{e}_1(v^2)$ are C^2-smooth unit vectors of the generatrices. By (6.13), the vector $\dot{\mathbf{e}}_1$ belongs to the tangent plane $T_P(F^2)$, being simultaneously orthogonal to e_1. Therefore,

$$\begin{aligned}
\dot{\mathbf{e}}_1 &= -k_1 \mathbf{e}_2 \\
\dot{\mathbf{e}}_2 &= k_1 \mathbf{e}_1 + k_2 \dot{\mathbf{e}}_3,
\end{aligned}$$

where k_1 and k_2 are, respectively, the geodesic and normal curvatures of the curve γ, and $\mathbf{e}_2$ is its unit tangent vector, e_3 is some unit vector in the normal plane $T_P^\perp(F^2)$ (the opposite signs before k_1 emerged as a consequence of the equality to zero of the product $(\mathbf{e}_1 \mathbf{e}_2)$ and its derivative). By the values of the geodesic curvature $k_1(P)$ and its derivative, we introduce the following classification of points of the directing curve γ:

$$\begin{aligned}
point\ P\ &-\ cylindrical,\quad &if\ k_1(P) = 0; \\
point\ P\ &-\ conical,\quad &if\ k_1(P) \neq 0,\ but\ \dot{k}_1(P) = 0; \\
point\ P\ &-\ torsial,\quad &if\ k_1(P) \neq 0,\ \dot{k}_1(P) \neq 0.
\end{aligned}$$

Now we will show that this classification makes it possible to split all two-dimensional developable surfaces into three respective classes. For this, we will investigate the geometry of the cuspidal edge Γ of the surface, which is determined by the following property: generatrices along it are its tangents, i.e., Γ is the envelope of the family of generatrices. Let us denote the position vector of the curve Γ as R and write down its representation as of some set of points of the surface F^2:

$$\mathbf{R}(v^2) = \boldsymbol{\rho}(v^2) + v^1(v^2)\mathbf{e}_1(v^2).$$

Calculate the tangent vector to Γ:

$$\dot{\mathbf{R}}(v^2) = \dot{v}^1\mathbf{e}_1 + (1 - k_1 v^1)\mathbf{e}_2,$$

therefore, it should be

$$k_1 v^1 = 1. \qquad (6.14)$$

Consider the possible cases.

1) At all points, $k_1(P) = 0$. Then there is no cuspidal edge on the surface, and $\dot{e}_1 = 0$, i.e., $e_1 = const$ – we have a 1-cylinder.

2) $k_1 \neq 0$ everywhere on all the surface (we recall that all considerations are local). Then by the equation (6.14) the cuspidal edge is determined by the unique way with $v^1 = (k_1)^{-1}$, which yields

$$\mathbf{R}(v^2) = \rho + (k_1)^{-1}\mathbf{e}_1, \quad \dot{\mathbf{R}}(v^2) = -(k_1)^{-2}\dot{k}_1\mathbf{e}_1.$$

Again there are two possibilities:

a) $\dot{k}_1 \equiv 0$; then the cuspidal edge degenerates into one point and we have that all generatrices emanate from this point – we have the case of a 1-conical surface.

b) $\dot{k}_1 \neq 0$ – the cuspidal edge really exists as some regular C^1-smooth curve and we have the case of a torse.

Further, it is easy to show that each generatrix consists of points of one type (this means that if we take an arbitrary point of a generatrix and draw its orthogonal directrix through this point, then the classification of the point using this directrix will be the same as it was in the respective point at the initial directrix Γ). Therewith, all points of the surface will be split into three non-intersecting classes, and the set of interior points of each class is either empty or makes an open set on the surface, into which set the generatrix of the surface as an open segment of a straight line can enter only totally. This means that the interior of each class is either a cylinder or a cone, or a torse, and different classes can border only along the generatrices[52].

Let us now pass on to the study of multidimensional developable surfaces $F^l \subset R^n$ of smoothness C^{l+1}. By Theorem 6.4 they are

---

[52] This description of the structure of a developable surface F^2 in $R^n, n \geq 3$, was obtained in the assumption of at least the C^3-smoothness of the surface and of the C^1-smoothness of its cuspidal edge. Recall that a similar description of the structure of a developable surface in R^3 was given in Subsection 3.2.1 in the assumption of only the C^2-smoothness of the surface.

$(l-1)$-ruled. Let us draw through an arbitrary point P of the considered surface F^l the trajectory orthogonal to the generatrix $(l-1)$-dimensional planes; we will obtain some C^{l+1}-smooth curve γ. We parametrize it by the natural parameter v^l, with designation of its position vector as $\rho(v^l)$ and let $e_l = \dot{\rho}(v^l)$. We introduce mutually orthogonal unit vectors $e_1, \ldots, e_{l-1}, e_l$ by the following recurrent ratios

$$
\begin{aligned}
\dot{\mathbf{e}}_l &= -k_{l-1}\mathbf{e}_{l-1} + k_l\mathbf{e}_{l+1} \\
\dot{\mathbf{e}}_{l-1} &= -k_{l-2}\mathbf{e}_{l-2} + k_{l-1}\mathbf{e}_l \\
\cdots\cdots & \cdots\cdots\cdots\cdots\cdots \\
\dot{\mathbf{e}}_i &= -k_{i-1}\mathbf{e}_{i-1} + k_i\mathbf{e}_{i+1} \\
\cdots\cdots & \cdots\cdots\cdots\cdots\cdots \\
\dot{\mathbf{e}}_{a+2} &= -k_{a+1}\mathbf{e}_{a+1} + k_{a+2}\mathbf{e}_{a+3} \\
\dot{\mathbf{e}}_{a+1} &= k_{a+1}\mathbf{e}_{a+2} \\
\dot{\mathbf{e}}_1 &= \dot{\mathbf{e}}_2 = \cdots = \dot{\mathbf{e}}_a = 0.
\end{aligned}
\tag{6.15}
$$

The sequence of these vectors is constructed as follows. The field of the unit vectors $\mathbf{e}_l$ tangent to γ is already known. The derivative of this field decomposes uniquely into its projections on $T_P(F^l)$ and on the normal plane. The normal component will make the vector $k_l(P)\mathbf{e}_{l+1}$, the horizontal component is designated as $-k_{l-1}(P)\mathbf{e}_{l-1}$; herewith, the vector $\mathbf{e}_{l-1}(P)$ is chosen such that the coefficient $k_{l-1}(P)$ be positive. As a $(l-1)$-dimensional plane generatrix, being the relative nullity space $T_P^\circ(F^l)$ of the surface, passes through the point P, the vector $\mathbf{e}_{l-1} \in T_P^\circ(F^l)$, and its derivative $\dot{\mathbf{e}}_{l-1}$, according to (6.13), will be assigned to $T_P(F^l)$. Then it will be decomposed into the sum of its orthogonal projections; $h\mathbf{e}_l$, into the direction of the vector $e_l \in T_P$; and $-k_{l-2}\mathbf{e}_{l-2} \in T_P^\circ$, into the orthogonal complement to e_l in T_P. Therefore, $\mathbf{e}_{l-2} \in T_P^\circ$, and then again due to (6.13) we have $\dot{\mathbf{e}}_{l-2} \in T_P(F^l)$ and we can repeat the previous constructions, remarking preliminarily that in view of $(\dot{\mathbf{e}}_l\mathbf{e}_{l-1}) + (\mathbf{e}_l\dot{\mathbf{e}}_{l-1}) = 0$ we have $h = k_{l-1}$ (the coefficient $k_{l-2}(P)$ is considered positive again). This reasoning on the consecutive definition of the vectors $\mathbf{e}_i$ is possible until at some step under a number $b \geq 1$ it proves that $k_{l-b}(P) = 0$, and we will not be able to define the vector $\mathbf{e}_{l-b}(P)$. Let $a = l - b$. If $k_a(P) = 0$, but P is the endpoint of some open arc on γ, at which $k_i \neq 0, a \leq i \leq l-1$, then we conduct the same constructions in the neighbourhood of any point P' from the same arc. But if $k_a(P) \equiv 0$ on all the curve γ or even in some neighbourhood of the point P, we choose at the point P as $\mathbf{e}_a, \mathbf{e}_{a-1}, \ldots, \mathbf{e}_1$ any vectors yielding in T_P an orthonormal basis together with the already available vectors $\mathbf{e}_l, \mathbf{e}_{l-1}, \ldots, \mathbf{e}_{a+1}$ and extend them along γ as constants in $E^n \supset F^l$ vectors. Let us make sure that thus constructed field of vectors $\mathbf{e}_1, \mathbf{e}_2, \ldots, \mathbf{e}_l$ will be an orthonormal basis in $T_P(F^l)$ along the entire arc of the curve γ, on which $k_{l-1} \neq 0, \ldots, k_{a+1} \neq 0$. For this, it is sufficient to show that the space $R^a(Q) = span\ (e_a(Q), \ldots, e_1(Q))$ at all points Q of the considered neighbourhood of the point P on γ is orthogonal to $R^{n-a}(Q) = span\ (\mathbf{e}_{a+1}(Q), \ldots, \mathbf{e}_l(Q), T_Q^\perp(F^l))$. This is true at the point P proper. Consider in $R^{n-a}(Q)$ an arbitrary vector field $\mathbf{f}(v^l)$ and make sure that its derivative along γ remains in $R^{n-a}(Q)$. Indeed, we have the decomposition

$$\mathbf{f}(v^l) = c^j(v^l)\mathbf{e}_j(v^l) + \mathbf{n}(v^l), a + 1 \leq j \leq l,$$

and $\mathbf{n}(v^l)$ is some field of vectors in the normal bundle $T_Q^\perp(F^l)$ along γ. Let $\mathbf{g}(v^l) \in T_Q^\circ(F^l)$ be an arbitrary smooth field of vectors on the surface along γ. By (6.13), its derivative $\dot{\mathbf{g}} \in T_Q(F^l)$, so $0 = \langle \dot{\mathbf{n}}(v^l)\mathbf{g}(v^l)\rangle + \langle \mathbf{n}\dot{\mathbf{g}}(v^l)\rangle = \langle \dot{\mathbf{n}}(v^l)\mathbf{g}(v^l)\rangle$. This means that $\dot{\mathbf{n}}(Q) \in span\ (\mathbf{e}_l(Q), T_Q^\perp) \subset span\ (\mathbf{e}_{a+1}(Q), \ldots, \mathbf{e}_l(Q), T_Q^\perp)$. Hence, we have

$$\dot{\mathbf{f}}(v^l) = \dot{c}^j(v^l)\mathbf{e}_j(v^l) + c^j(v^l)\dot{\mathbf{e}}_j(v^l) + \dot{\mathbf{n}}(v^l) \in$$
$$span\ (\mathbf{e}_{a+1}(Q), \ldots, \mathbf{e}_l(Q), T_Q^\perp(F^l)).$$

Therefore, the space $R^{n-a}(Q)$ remains stationary along γ, and as at the point P it is orthogonal to the likewise stationary space $R^a(Q) = span\ (\mathbf{e}_a(Q), \ldots, \mathbf{e}_1(Q))$, then the union of their bases is an orthonormal base in $R^n(Q)$ along the curve γ. This means that we can represent the position vector of the surface F^l in the following form

$$\mathbf{r}(v^1, \ldots, v^l) = \boldsymbol{\rho}(v^l) + v^i\mathbf{e}_i(v^l), 1 \leq i \leq l - 1.$$

By analogy with the case of two-dimensional surfaces, we now also need to introduce some classification of points of the surface F^l. For this, let us introduce the concept of the *cylindroconicity index* (a, b). Here, an integer a – the cylindricity index – is the same number a, which was defined when deducing the relations (6.15). Therefore, the stationary space R^a with the basis of constant vectors $e_1, \ldots, e_a$ can be called the *direction of cylindricity* of the surface F^l. The definition of the conicity index $b = 0$ or $b = 1$ is more complicated and could not be written out in the general case exactly. Let us explain the algorithm of its calculation. We will search for the cuspidal edge by analogy with two-dimensional surfaces in the form of a curve

$$\mathbf{R}(v^l) = \boldsymbol{\rho}(v^l) + v^i(v^l)\mathbf{e}_i(v^l), \quad 1 \leq i \leq l - 1,$$

using the condition of its tangency with the generatrices coming to its points. But this time the generatrices are not straight lines but planes, so we assume that the cuspidal edge is tangent to the last (by the order of construction) non-constant direction e_{a+1}. With consideration for the relations (6.15), we have

$$
\begin{aligned}
\dot{\mathbf{R}}(v^l) \;=\; & \dot{\boldsymbol{\rho}} + \dot{v}^i \mathbf{e}_i + v^i \dot{\mathbf{e}}_i = \\
& \dot{v}^1 \mathbf{e}_1 + \cdots + \dot{v}^a \mathbf{e}_a + \\
& (\dot{v}^{a+1} - v^{a+2} k_{a+1})\mathbf{e}_{a+1} + \\
& (\dot{v}^{a+2} + v^{a+1} k_{a+1} - v^{a+3} k_{a+2})\mathbf{e}_{a+2} + \\
& \ldots\ldots + \\
& (\dot{v}^i + v^{i-1} k_{i-1} - v^{i+1} k_i)\mathbf{e}_i + \\
& \ldots\ldots + \\
& +(\dot{v}^{l-1} + v^{l-2} k_{l-2})\mathbf{e}_{l-1} + \\
& +(1 + v^{l-1} k_{l-1})\mathbf{e}_l.
\end{aligned}
$$

To have $\dot{\mathbf{R}} \parallel \mathbf{e}_{a+1}$ for v^l in some range of its change, it is required to demand the satisfaction of the equalities $\dot{v}^1 = \cdots = \dot{v}^a = 0$ and all coefficients at $\mathbf{e}_i$, $a + 2 \leq i \leq l$. This condition gives the following system of equations:

$$
\begin{aligned}
v^{l-1} k_{l-1} + 1 \;&= 0 \\
v^{l-2} k_{l-2} \;&= -\dot{v}^{l-1} \\
&\ \ldots\ldots \\
v^{i-1} k_{i-1} \;&= -\dot{v}^i + v^{i+1} k_i \\
&\ \ldots\ldots \\
v^{a+1} k_{a+1} \;&= -\dot{v}^{a+2} + v^{a+3} k_{a+2}
\end{aligned}
\qquad (6.16)
$$

We can solve this equation consecutively, starting from the first one:

$$v^{l-1} = f_{l-1}(k_{l-1}) \equiv -(k_{l-1}(v^l))^{-1}$$
$$v^{l-2} = f_{l-2}(k_{l-1}, k_{l-2}, \dot{k}_{l-1}) \equiv -(k_{l-2}(v^l))^{-1}(\dot{k}_{l-1}(v^l))/k_{l-1}^2(v^l)$$
$$\cdots \qquad \cdots\cdots\cdots$$
$$v^{a+1} = f_{a+1}(k_{l-1}, k_{l-2}, \ldots, \dot{k}_{l-1}, \ddot{k}_{l-1}, \ldots).$$

When we substitute these values v^i, $a+1 \leq i \leq l-1$, into the equality $\dot{\mathbf{R}}(v^l) = (\dot{v}^{a+1} - v^{a+2}k_{a+1})\mathbf{e}_{a+1}$, we will obtain some expression of the form

$$\dot{\mathbf{R}}(v^l) = f(k_{l-1}, \ldots k_{a+1})\mathbf{e}_{a+1},$$

in which the function $f(k_{l-1}, \ldots k_{a+1})$ depends in the known way on the "curvatures" k_i, $a + 1 \leq i \leq l - 1$, and their derivatives up to some order. Well then, the conicity coefficient b in the cylindroconicity index (a, b) is equal to 1, if $f(k_{l-1}, \ldots k_{a+1}) = 0$, and is 0, if $f(k_{l-1}, \ldots k_{a+1}) \neq 0$. At $b = 1$, the cuspidal edge degenerates into a point, and we do have an analogue of the conical surface. In a particular case $a = l - 2$ the condition f has the form $f = k_{l-1} = 0$.

Further, it can be shown that on any fixed generatrix the cylindroconicity index is constant. This observation makes it possible to prove that a set of *stable* points of a surface with the cylindroconicity index (a, b), which have an open neighbourhood of points with the same index, is an everywhere dense and open set in surface topology.

Finally, everything is ready for describing the structure of developable multidimensional surfaces $F^l \subset R^n$.

Theorem 6.7. *Any C^{l+1}-smooth developable surface $F^l \subset R^n$ in the neighbourhood of a stable point is either a torse or a cylindrocone over some torse. (See [179].)*

And in the general case a C^{l+1}-smooth developable surface is a union of closure of the domains, each of which is either a torse or cylindrocone over a torse.

Proof of the theorem. Let $U \subset F^l$ be some neighbourhood of a point P with the constant index of cylidroconicity (a, b). Consider the surface $F^{l-a} \subset R^{n-a}$, obtained in the intersection of the surface F^l with the plane E^{n-a}, passing through the point P orthogonally to the direction of the cylindricity R^a. The surface F^{l-a} is a surface of rank 1 in its space R^{n-a} and has a cylindroconicity index $(0, b)$. Two cases are possible relative to the conicity index b.

1) $b = 0$. In this case, the surface F^{l-a} has a cuspidal edge and, therefore, is a $(l - a)$-torse, but then the entire surface F^l will be an a-cylinder over the $(l - a)$-torse F^{l-a}.

2) $b = 1$. In this case, for the surface F^{l-a} the cuspidal edge degenerates into some point $A \neq P$ and, thus, it is a 1-cone with the vertex at A. Consider the surface $F^{l-a-1} \subset R^{n-a-1}$, obtained in the intersection of the surface F^{l-a} by the plane E^{n-a-1} passing through the point P orthogonally to the direction AP. A new surface F^{l-a-1} then will be a surface of rank 1 in its space and its cylindroconicity index will be $(0, 0)$. This means that the surface F^{l-a-1} is an $(l - a - 1)$-torse, and all the surface will be an $(a + 1)$-cylindrocone over it.

6.2   *Isometric immersions of complete multidimensional locally Euclidean manifolds*

At local immersions of l. E. metrics we had to impose an *a priori* requirement on the structure of the surface, so that its nullity index have a maximal possible value (if, certainly, we search for surfaces distinct from the plane). But when considering immersions of complete, in particular, compact metrics, an important role is already played by the dimensions of the manifold and of the space, into which we want to immerse it, and the existence of plane generatrices at some relations between dimensions emerges by itself, without any *a priori* assumption of their occurrence.

The first result in this round of issues was the following theorem:

Theorem 6.8. *A compact n-dimensional manifold with a locally Euclidean metric could not be isometrically and C^2-smoothly immersed into a Euclidean space of dimension less than $2n$. (See [175].)*

The proof is based on the fact that on a compact surface without the boundary there is a point, at which the scalar square r^2 of the position vector of the surface reaches its maximum[53]. Then, using

---

[53] The work [175] itself discusses the impossibility of *embedding*, but, apparently, the reasoning associated with the existence of an r^2 maximum is valid for immersions, too.

Frenet decompositions, i.e., decompositions of the form

$$\frac{\partial \mathbf{p}_i}{\partial u^j} = \Gamma_{ij}^k \mathbf{p}_k + L_{ij}^\alpha \mathbf{t}_\alpha,$$

where $\mathbf{p}_i = \dfrac{\partial \mathbf{r}}{\partial u^i}, 1 \leq i \leq n$, and $\mathbf{t}_\alpha$ are $(n-1)$ mutually orthogonal unit normals, with account for the Gaussian equations

$$L_{ij}^\alpha L_{km}^\alpha = L_{im}^\alpha L_{kj}^\alpha$$

on the basis of some algebraic lemma it is established that the required conditions of the maximum are not satisfied.

In [171], the proposition of the theorem is expanded to immersions of a compact manifold with an l. E. metric in the space of constant negative curvature.

Further results on some *a priori* properties of isometric immersions of multidimensional Euclidean metrics were obtained in [116].

Theorem 6.9. *Let $\psi : M^d \to E^{d+k}$ be some isometric immersion of a d-dimensional complete flat Riemannian manifold into Euclidean space. Then M^d contains a totally geodesic submanifold, which is mapped isometrically onto some n-dimensional plane in E^{d+k}, where n is equal to the minimal value of the nullity index of the mapping ψ.*

According to [38], the dimensions of the surface l, space N and the value of the nullity index ν are related by the inequality $\nu \geq 2l - N$. In the case of the theorem, this gives the condition $n \geq d - k$, so it yields such a corollary:

Corollary 6.4. *If in the conditions of the theorem we have $k < d$, then the surface $\psi(M^d)$ contains a plane of dimension $d - k$.*

At some additional condition for the structure of the surface, we can propose the existence on it of not only one plane but of the cylindricity of the surface in the large. Namely, the same work [116] has proved that a complete surface $\psi : M^d \to E^{d+k}$ with an l. E. metric is an n-cylinder if: 1) the nullity index has a constant value n and 2) the surface has the flat normal connection. In [170] this result is proved already without the condition 2).

These propositions were refined in [75], which proves the following theorem:

Theorem 6.10. *Let an d-dimensional $C^k (k \geq 2)$-smooth Riemannian manifold[54] M^d with a complete l. E. metric be immersed isometrically into Euclidean space E^{d+e}, $e < d$ by the mapping $f : M^d \to E^{d+e}$ of smoothness C^k. Then the minimal value m of the nullity index $\nu(p)$ of points p of the surface $f(M^d)$ is positive and the surface is m-cylindrical over some C^k-smooth surface S of dimension $d - m$. What is more, the very manifold M^d proves to be split into the product $E^m \times M^{d-m}$, where M^{d-m} is some complete Riemannian manifold of smoothness C^k, and the mapping f is represented, respectively, as $f = id \times g$ with $g(M^{d-m}) = S$.*

The order of cylindricity is exact, in the sense that in the large the surface cannot be $(m + 1)$-cylindrical. Besides, as the number $m = \min \nu(p), p \in f(M^d)$ satisfies the inequality $m \geq d - e$, then the order of cylindricity of the surface $f(M^d)$ is equal to at least $d - e$. In particular, hypersurfaces (case $e = 1$) are $(d - 1)$ cylinders over the curve (the earlier mentioned result from [76]). Note that in the case of smoothness of class C^2 (i.e., when $k = 2$ and the metric is of class C^1) the existence of a continuous tensor of curvature is guaranteed by the C^2 smoothness of the isometric immersion, and the actual $W_2^p (p > d)$ smoothness of the metric (the existence of second order derivatives of class L_p in the Sobolev sense) follows from [2]. Obviously, Corollary 6.4 and Theorem 6.10 are generalizations of Theorem 6.8, as the presence of a plane in the surface makes its boundedness impossible.

We can see that Theorem 6.10 at some relation of dimensionalities automatically guarantees the ruled structure of the surface with a complete l. E. metric.

As for non-complete multidimensional l. E. metrics, we are unaware of any special theorems of their immersions. In particular, still unsolved is the simplest **question** formulated in [66] (and also repeated in [26]):

Open question 6.2. Into which minimal-dimension Euclidean space is any l. E. metric given in a closed n-dimensional ball $(n \geq 3)$ embedded?•

---

[54] This means that a metric tensor of smoothness C^{k-1} is given on a manifold of smoothness C^k.

We would also like to include into the number of unsolved problems the analogue of the Plateau problem with the substitution of the condition $H = 0$ by the condition $K = 0$:

Open question 6.3. If a closed contour in E^3 is given, under which conditions can we span a developable surface over it, and how such a surface can be found if it does exist?$\bullet$

In conclusion, we would like to note the existence of vast literature on bendings and infinitesimal bendings of developable surfaces. The first global result on the subject appears to belong to Vilkitsky (1936), who proved that the complete cylinder with a closed directrix (in reality, even a part of a cylinder unbounded in one direction and cut out by some closed curve) admits bendings[55] only with the generatrices preserved[56]. This proposition without proof is mentioned at the very end of the famous paper by Cohn-Vossen [43] (reprinted in [44])[57]. As I had no chance to come across the proof or even mention of the result by Vilkitsky in the literature, I consider it appropriate to present here the proof of his really beautiful proposition. Let S be a complete doubly connected cylindrical surface, and S_1 be a developable surface of usual ruled geometry isometric to it (e.g., a C^1-smooth torse). Then S_1 is also a cylinder (on the strength of Theorem 3.3 at the C^1 smoothness of S_1, or of Theorem 3.4, if we assume that at least one complete straight line lies on S_1; without these theorems proved much later after Vilkitsky, the cylindricity of S_1 can be established based on the following *observation*: if two non-parallel geodesics pass on a cylinder (in the intrinsic sense) through

---

[55] As we will see from the proof, the case in hand is in fact isometric transformations both in continuous and discrete sense.

[56] It is worth noting here that in the second part of the known monograph by V.F. Kagan ([95], §70) it is asserted that in bendings the property of the transition of generatrices into generatrices is valid even locally for any ruled surfaces, with the possible exception of the case of isometry on a second-order surface. Using examples, it is easy to show that for developable surfaces this property of generatrices' preservation in bendings is locally false for each of the three classes of zero-curvature surfaces, so that Vilkitsky's result is indeed a fact of geometry in the large.

[57] We are unaware of any information about Vilkitsky besides that reported in [43]: at that time he was a student of Leningrad University.

points of some of its closed geodesic G (i.e., a segment perpendicular to the edges of the development of the cylinder in the form of an infinite band Π between two parallel lines on the plane), then during their unbounded extension to the same side from G their extensions would intersect). Let us intersect S by some plane orthogonal to the generatrices. In the intersection, we obtain a closed geodesic G. By the isometry, on the cylinder S_1 it has some corresponding geodesic G_1. Take on G_1 an arbitrary point M and draw through it the section of the surface S_1 by the plane orthogonal to its generatrices. In the section, we obtain a closed geodesic curve G'. As only one closed geodesic can be drawn through the given point on the cylinder, the geodesics G_1 and G' coincide. Therefore, the pre-images of the generatrices both from S and from S_1 on the band are orthogonal to the same segment – the common pre-image of the geodesics G and G_1, so the isometry between S and S_1 preserves the generatrices. In the case of a "half-cylinder", we cannot refer to either of Theorems 3.3 and 3.4, but then to establish the cylindricity of S_1 we need to be based on the above presented observation on the intersection of non-parallel geodesics and, on the assumption of the presence of domains with planar points on S_1, on the geometry of such domains. Besides, we need to assume that no plane "tongues", which could be cut by the rectilinear segment without disturbing the double connection of S, are adjoined to the edge of the "half-cylinder" S (otherwise the bending could have been performed only within the limits of those "tongues"; see [173], p. 426 about a similar situation; evidently, in order to avoid such a situation, it is sufficient to cut the cylinder by a closed geodesic).

We are unable to list all the major works on the bendability/non-bendability and infinitessimal rigidity/nonrigidity of developable surfaces. We would mention only the papers [49, 151, 173], in which these problems are considered in a global statement. The bendability/nonbendability of tori in S^3 is discussed in [57, 98, 100] (generalization [57] by dimension), [101, 146] and [188]. A fresh topic in the problems of bendings of developable surfaces is proposed in [162, 164, 165]. The reviews [88–90] provide more detailed information on the works on isometric deformations.

Acknowledgements

I am pleased to use this occasion to thank my Western colleagues H. Stachel and J. Wallner (Austria), José A. Gálvez and Pablo Mira (Spain), H. Rosenberg (France), D. Hoffman (USA) as well as my Russian colleagues V.A. Alexandrov and S.V. Markelov for help in literature search. I also thank Yu.D. Burago for useful consultations, D.I. Sabitov for assistance in artwork preparation and of course Victor Selivanov, who translated my work into English in the most thorough and responsible way.

References

1. Alexandrov A.D., *Intrinsic Geometry of Convex Surfaces*, Moscow: Gostekhizdat, 1948 (German translation: *Die innere Geometrie der konvexen Flächen*, Berlin: Akademie Verlag, 1955; English translation: Intrinsic Geometry of Convex Surfaces, Boca Raton: Chapman and Hall/CRC, Taylor and Francis Group, 2006).   (pp. 3, 50)

2. Alexandrov A.D., Berestovski V.N. and Nikolaev I.G., Generalized Riemannian Spaces, *Usp. Mat. Nauk*, 1986, vol. 43, No 3, pp. 3–44.   (pp. 25, 223, 250)

3. Alexandrov A.D. and Zalgaller V.A., Twodimensional Manifolds of Bounded Curvature, *Trudy Mat. Inst. AN SSSR imeni Steklova*, 1962, vol. 63, pp. 3–262.   (pp. 28, 70)

4. Alexandrov V.A., Embedding of Locally Euclidean and Conformal Metrics, *Mat. Sbornik*, 1991, vol. 182, No 2, pp. 1105–1117.   (pp. 223, 225)

5. Aledo J., Gálvez J. and Mira P., A D'Alembert Formula for Flat Surfaces in the 3-Sphere, *J. Geom. Anal.*, 2009, vol. 19, pp. 211–232.   (pp. 98, 196, 203)

6. Aminov Yu.A., Differential Geometry and Topology of Curves, Moscow: Fizmatgiz, 1978.   (p. 63)

7. Aminov Yu.A., Problems of Embedding: Geometrical and Topological Aspects, *Itogi Nauki i Tekhniki. Problemy Geometrii*, vol. 23, Moscow: VINITI, 1982, pp. 119–156.   (p. 3)

8.   Aminov Yu.A., *Geometry of Submanifolds*, Kiev: Naukova
     Dumka, 2002.   (pp. 3, 182, 191)

9.   Aminov Yu.A., Physical Interpretation of Some Ruled surfaces
     in E^3 in Terms of the Motion of a Point Charge, *Matem. Sbornik*,
     2006, vol. 197, No 12, pp. 3–10.   (p. 108)

10. Avkhadiev F.G. and Aksent'ev L.G., The Main Results in Suf-
     ficient Conditions for Univalence of Functions, *Usp. Mat. Nauk*,
     1975, vol. 30, No 4, pp. 3–60.   (p. 53)

11. Bakelman I.Ya., Chebyshev Networks in Manifolds of Bounded
     Curvature, *Trudy Mat. Inst. AN SSSR imeni Steklova*, 1965,
     vol. 76, pp. 124–129.   (p. 38)

12. Bari N.K., *Trigonometrical Series*, Moscow: Fizmatgiz, 1959.
                                                                (p. 11)

13. Barr S., *Experiments in Topology*, Crowell: New York, 1964.
                                                               (p. 177)

14. Bazilevich I.E., Domains of Initial Coefficients of Bounded Uni-
     valent Functions with n-tiple Symmetry, *Mat. Sbornik*, 1957,
     vol. 43, No 4, pp. 409–428.   (p. 55)

15. Berestovski V.N. and Nikolaev I.G., Multidimensional Gener-
     alized Riemannian Spaces, *Itogi Nauki i Tekhniki. Sovr. Prob-
     lemy Matematiki. Fund. Napravleniya*, vol. 70, Geometriya-4,
     Moscow: VINITI, 1989, pp. 190–272. (English translation: *En-
     cycl. Math. Sci.*, 70, Berlin: Springer, 1993, pp. 168–243).
                                                          (pp. 25, 223)

16. Bers L., John F. and Schechter M., *Partial Differential Equa-
     tions*, Interscience Publishers: New York–London–Sydney, 1964.
                                                                (p. 26)

17. Bianchi L., Sulle superficie a curvatura nulla negli spazi a cur-
     vatura costante. *Atti Acc. delle Scienze di Torino*, 1895, 30, pp.
     475–487 (see also in L. Bianchi, Opere, vol. VIII, Roma, 1958,
     pp. 256–265).   (pp. 179, 184, 191, 200)

18. Bianchi L., Sulle superficie a curvatura nulla in geometria ellittica, *Ann. di mat. pura e appl.*, 1896, 24(2), pp. 93–129 (see also in L. Bianchi, Opere, vol. VIIl, Roma, 1958, pp. 266–301).
(pp. 179, 187, 191, 194, 200)

19. Biernacki M., Sur les fonctions univalents, *Matematica (Cluj)*, 1936, 12, pp. 49–64.
(p. 52)

20. Blanuša B., Le plongement isometrique de la bande de Moebius infiniment large euclidienne dans un espace a quatre dimensions, *Bull. Int. Acad. Journ. Nouvelle Serie*, 1954, livre 12 (Classe de Sc. Math. 4), pp. 19–23.
(pp. 65, 204, 205)

21. Bojarski B., On the Beltrami Equation Once Again: 54 Years Later (to appear in *Ann. Acad. Sci. Fenn. Math.*, 2009).
(p. 26)

22. Bolza O., *Vorlesungen über Variationrechnung*, Leipzig und Berlin, 1909.
(p. 6)

23. Borisenko A.A., On the Structure of Continuous Surfaces Containing a Straight Line, *Ukr. Geom. Sbornik*, 1973, 14, pp. 21–24.
(p. 91)

24. Borisenko A.A., Extrinsic Geometry of Parabolic and Saddle Multidimensional Submanifolds, *Usp. Mat. Nauk*, 1998, 53, No 6, pp. 3–52.
(pp. 3, 179)

25. Borisenko A.A., Isometric Immersions of Space Forms in Riemannian and pseudo-Riemannian Spaces of Constant Curvature, *Usp. Mat. Nauk*, 2001, 56, No 3, pp. 3–78.
(pp. 3, 196)

26. Borisenko A.A., *Intrinsic and Extrinsic Geometry of Multidimensional Submanifolds*, Moscow: Examen, 2003.
(pp. 3, 91, 179, 184, 196, 203, 216, 236, 250)

27. Borisov Yu.F., Parallel Transportation on a Smooth Surface, *Vestn. Lenigr. Univ. Mal. Mekh. Astr.*, 1958, No 7:2, pp. 160–171; 1958, No 19:3, pp. 45–54; 1959, No 1:1, pp. 34–45; 1959, No 14:3, pp. 83–92; corrections in 1960, No 19:4, pp. 127–129.
(p. 93)

28. Borisov Yu.F., $C^{1,\alpha}$-Isometric Immersions of Riemannian Spaces, *Dokl. Akad. Nauk SSSR*, 1965, vol. 163, No 1, pp. 11–13.
(p. 72)

29. Borisov Yu.F., Nonregular Surfaces of Class $C^{1,\beta}$ with Analytic Metrics, *Sib. Math. J.*, 2004, vol. 45, No 1, pp. 25–61.
(pp. 72, 93, 239)

30. Browder F., Nonlinear Operators and Nonlinear Equations of Evolution in Banach Spaces, *Proc. Symposia in Pure Math.*, 1976, 18, part 2, Providence: Amer. Math. Soc.   (p. 228)

31. Burago Yu.D., Inequalities of Isoperimetric Type in the Theory of Surfaces of Bounded Exterior Curvature, *Zapiski Nauchn. Seminarov LOMI*, 1968, vol. 10, pp. 3–203.   (p. 81)

32. Burago Yu.D., Geometry of Surfaces in Euclidean Spaces, *Itogi Nauki i Tekhniki, Sovr. Problemy Matematiki. Fund. Napravleniya*, vol. 48, Geometriya-3, Moscow: VINITI, 1989, pp. 5–97 (English translation: *Encycl. Math. Sci.*, 48, Berlin: Springer, 1992, pp. 1–85).   (pp. 3, 76, 81, 82)

33. Burago Yu.D. and Zalgaller V.A., Isometric Piecewise Immersions of Two-dimensional Manifolds with Polyhedral Metric in R^3, *Algebra i analyse*, 1995, vol. 7:3, pp. 76–95. (pp. 177, 195)

34. Bushmelev A.V., Isometric Embeddings of an Infinite Flat Moebius Band and a Flat Klein Bottle into E^4, *Vestn. Mosk. Univ. Mat. Mekh.*, 1988, No 3, pp. 38–41.
(pp. 207, 208, 209, 214, 217)

35. Calabi E. and Hartman Ph., On the Smoothness of Isometries, *Duke Math. J.*, 1970, vol. 37, No 4, pp. 741–750.
(pp. 4, 5, 19, 20, 21, 195)

36. Charlap L., *Bieberbach Groups and Flat Manifolds*, New York: Springer, 1982.   (p. 143)

37. Chakmazyan A.V., Submanifolds in Constant Curvature Spaces with a Parallel Subbundle of the Normal Bundle, *Ukr. Geom. Sbornik*, 1977, vol. 20, pp. 132–140.   (p. 182)

38. Chern S.S. and Kuiper N.H., Some Theorems on the Isometric Imbedding of Compact Riemann Manifolds in Euclidean Space, *Ann. Math.*, 1952, vol. 56, pp. 422–430. (pp. 150, 162, 167, 249)

39. Chern S.-S. and Osserman R., Remarks on the Riemannian Metric of a Minimal Submanifold, *Lecture Notes in Math.*, 1981, vol. 894, pp. 49–90.   (p. 144)

40. Chen X.-X. and Peng C.-K., Deformation of Surfaces Preserving Principal Curvatures, *Lecture Notes of Math.*, 1989, vol. 1369, pp. 63–70.   (p. 144)

41. Chen B.-Y. and Yano K., Submanifolds Umbilical with Respect to a Non-parallel Normal Subbundle, *Kodai Math. Semin. Reports*, 1973, vol. 25, No 3, pp. 289–296.   (p. 182)

42. Chicone C. and Kalton N.J., Flat Embeddings of the Möbius Strip in R^3, *Comm. Appl. Nonlinear Anal.*, 2002, vol. 9, pp. 31–50.   (p. 197)

43. Cohn-Vossen S., Bendability of Surfaces in the Large, *Usp. Mat. Nauk*, 1936, vol. 1, pp. 33–75.   (p. 251)

44. Cohn-Vossen S., *Some Questions of Differential Geometry in the Large*, Moscow: Fizmatgiz, 1959.   (pp. 144, 251)

45. Dadok J. and Sha J., On Embedded Surfaces in S^3, *J. Geom. Anal.*, 1997, vol. 7, No 1, pp. 47–55.   (pp. 179, 201, 202, 203)

46. Darboux G., *Léçons sur la théorie générale des surfaces*, vol. 3, Paris, 1894.   (p. 36)

47. Dajczer M. *et al.*, Submanifolds and Isometric Immersions, *Math. Lecture Series*, 13, Houston: Publish or Perish Inc., 1990.   (pp. 179, 182)

48. Dajczer M. and Florit L., Compositions of Isometric Immersions in Higher Codimension, *Manuscripta Math.*, 2001, vol. 105, pp. 505–517.   (p. 180)

49. Dajczer M. and Gromoll D., Rigidity of Complete Euclidean Hypersurfaces, *J. Diff. Geometry*, 1990, vol. 31, pp. 401–416.
(p. 252)

50. Dajczer M. and Tojeiro R., On Composition of Isometric Immersions, *J. Diff. Geometry*, 1992, 36, No 1, pp. 1–18. (p. 179)

51. Dajczer M. and Tojeiro R., Submanifolds with Nonparallel First Normal Bundle, *Can. Math. Bull.*, 1994, 37, No 3, pp. 330–337.
(pp. 179, 181, 184)

52. Dajczer M. and Tojeiro R., On Flat Surfaces in Space Forms, *Houston J. Math.*, 1995, 21, No 2, pp. 319–338. (pp. 179, 183)

53. do Carmo M. and Dajczer M., Local Isometric Immersions of R^2 into R^4, *J. reine und angewand. Math.*, 1993, 442, pp. 205–219.
(pp. 179, 184)

54. Efimov N.V., A Study of the Single-valued Projection of a Surface of Negative Curvature. *Dokl. Akad. Nauk SSSR*, 1953, vol. 93, No 4, pp. 609–611. (p. 97)

55. Enomoto K., The Gauss Map of Flat Surface in R^4, *Kodai Math. J.*, 1986, vol. 9, No 1, pp. 19–32. (pp. 179, 201)

56. Enomoto K., Global Properties of the Gauss Image of Flat Surfaces in R^4, *Kodai Math. J.*, 1987, vol. 10, pp. 272–284.
(pp. 179, 201)

57. Enomoto K., Kitagawa Y. and Wiener J., A Rigidity Theorem for the Clifford Tori in S^3, *Proc. Am. Math. Soc.*, 1996, 124, No 1, pp. 265–268. (pp. 198, 252)

58. Gakhov F.D., *Boundary Problems*, Moscow: Nauka, 1977.
(pp. 42, 62, 64)

59. Gálvez J.A., Martinez A. and Milan F., Flat Surfaces in the Hyperbolic 3-Space, *Math. Ann.*, 2000, vol. 316, pp. 419–435.
(p. 178)

60. Gálvez J.A., Martinez A. and Milan F., Flat Surfaces in L^4, *Ann. Glob. Anal. Geom.*, 2001, vol. 20, No 3, pp. 243–251.
(pp. 178)

61. Gálvez J.A. and Mira P., Isometric Immersions of R^2 into R^4 and Perturbation of Hopf Tori, arXiv:math. Dg/0301075 vl 9 Jan 2003, 4, pp. 1–26 (to appear in *Math. Z.*).
(pp. 179, 183, 184, 197, 203)

62. Gálvez J.A., Martinez A. and Mira P., The Space of Solutions to the Hessian One Equation in the Finitely Punctured Plane, *J. de Math. Pure Appl.*, 2005, vol. 84, No 12, pp. 1744–1757.
(p. 178)

63. Gluck H., Krigelman K. and Singer D., The Converse to the Gauss–Bonnet Theorem in PL, *J. Diff. Geometry*, 1974, 9, No 4, pp. 601–616.
(p. 144)

64. Goluzin G.M., *The Geometric Theory of Functions*, Moscow: Nauka, 1966.
(pp. 49, 51, 53)

65. Gromol D., Klingenberg W. and Meyer W., *Riemanische Geometrie in Grossen*, Berlin: Springer-Verlag, 1968.
(p. 20)

66. Gromov M., *Partial Differential Relations*, Berlin: Springer-Verlag, 1986.
(pp. 121, 217, 250)

67. Gao H. and Wu Z., On Beltrami Equation with Double Characteristic Matrices, *Acta Math. CSci.*, *Ser. A*, *Chin. Ed.*, 2002, vol. 22, No 4, pp. 433–440.
(p. 26)

68. Hadamard J., Sur les transformations ponctuelles, *Bull. Soc. Math. France*, 1906, 34, pp. 71–84.
(p. 228)

69. Han Q. and Hong J.-X., Isometric Embedding of Riemannian Manifolds in Euclidean Spaces, *AMS, Mathematical Surveys and Monographs*, vol. 130, 2006, pp. 1–260.
(pp. 3, 179, 216)

70. Halperin B. and Weaver C., Inverting a Cylinder through Isometric Immersions and Isometric Embeddings, *Trans. Amer. Math. Soc.*, 1977, vol. 230, pp. 41–70. (pp. 150, 151, 154, 176)

71. Hartman Ph., On Unsmooth Two-dimensional Riemannian Metrics, *Amer. J. Math.*, 1952, vol. 74, pp. 215–226.   (p. 23)

72. Hartman Ph., On Isometries and on a Theorem of Liouville, *Math. Zeitschr.*, 1958, Bd. 69, S. 202–210.   (pp. 23, 27)

73. Hartman Ph., *Ordinary Differential Equations*, New York: Wiley, 1964.   (p. 220)

74. Hartman Ph., On Isometric Immersion in Euclidean Space of Manifolds with Nonnegative Sectional Curvature, *Trans. Amer. Math. Soc.*, 1965, 115, No 2, pp. 94–109.   (p. 249)

75. Hartman Ph., On the Isometric Immersion in Euclidean Space of Manifolds with Nonnegative Sectional Curvature, *Trans. Amer. Math. Soc.*, 1970, 147, No 2, pp. 541–561.

    (pp. 225, 236, 249)

76. Hartman Ph. and Nirenberg L., On Spherical Image Maps whose Jacobians do not Change Sign, *Amer. J. Math.*, 1959, vol. 81, No 4, pp. 901–920.   (pp. 75, 82, 85, 109, 113, 179, 195, 250)

77. Hartman Ph. and Wintner A., On the Fundamental Equations of Differential Geometry, *Amer. J. Math.*, 1950, vol. 72, No 4, pp. 757–772.   (pp. 73, 74, 75, 95)

78. Hartman Ph. and Wintner A., Gaussian Curvature and Local Embedding, *Amer. J. Math.*, 1951, vol. 73, No 4, pp. 876–882.

    (p. 36)

79. Hartman Ph. and Wintner A., On the Problem of Geodesics in the Small, *Amer. J. Math.*, 1951, vol. 73, No 1, pp. 132–148.

    (pp. 6, 145)

80. Hartman Ph. and Wintner A., On the Asymptotic Curves of a Surface, *Amer. J. Math.*, 1951, 73, No 1, pp. 149–172.

    (pp. 86, 99. 107, 109)

81. Hartman Ph. and Wintner A., On the Solutions of Certain Overdetermined Systems of Partial Differential Equations, *Archiv Math.*, 1954, 5, pp. 168–174.   (p. 27)

82. Heinz E., Über Flächen mit eineindeutigen Projektion auf eine Ebene, deren Krümmungen durch Ungleichungen eingeschränkt sind, *Math. Ann.*, 1955, vol 129, No 5, pp. 451–454.   (p. 97)

83. Hilbert D., Über das Dirichlet'sche Princip, *Jahresbericht der Deutschen Mathematiker-Vereinigung*, vol. 8 (1900), S. 184–188.   (p. 6)

84. Hoffman D. and Karcher H., Complete Embedded Minimal Surfaces of Finite Total Curvature, in *Encyclopedia of Mathematics*, 1997, pp. 5–93; R. Osserman (ed.), Springer Verlag.   (p. 144)

85. Hoffman D., Weber M. and Wolf M., An Embedded Genus-one Helicoid, arXiv: math.DG/0401080 v2, 2004, pp. 1–115.   (p. 144)

86. Hsiung C.-C., *A First Course in Differential Geometry*, New York: Wiley, 1981.   (p. 175)

87. Ivanov A.B., Isometric Immersions of Complete Two-dimensional Locally Euclidean Metrics into Euclidean spaces, *Mat. Zametki*, 1973, vol. 13, No 3, pp. 497–499.   (p. 216)

88. Ivanova-Karatopraklieva I. and Sabitov I.Kh., Bendings of Surfaces. I, *Itogi Nauki i Tekhniki. Probl. Geometrii*, vol. 23, Moscow: VINITI, 1991, pp. 131–184 (English translation: *J. Math. Sci.* 1991, vol. 70, No 2, pp. 1685–1716).   (p. 252)

89. Ivanova-Karatopraklieva I. and Sabitov I.Kh., Bendings of Surfaces. II, *Itogi Nauki i Tekhniki. Sovr. Mat. Prilozh. Tematich. Obzory*, vol. 8, Moscow: VINITI, 1995, pp. 108–168 (English translation: *J. Math. Sci.* 1993, vol. 74, No 3, pp. 997–1043).   (p. 252)

90. Ivanova-Karatopraklieva I., Markov P.E. and Sabitov I.Kh., Bendings of Surfaces. III, *Fund. Prikl. Mat.*, 2006, 12, No 1, pp. 1–54.   (p. 252)

91. Jörgens K., Über die Lösungen der Differentialgleichung $rt - s^2 = 1$, *Math. Ann.*, 1954, vol. 127, No 2, pp. 130–134.   (p. 178)

92. Jörgens K., Harmonische Abbildungen und die Differential-
    gleichung $rt - s^2 = 1$, *Math. Ann.*, 1955, vol. 129, No 3, pp.
    330–344.   (p. 178)

93. John F., On Quasi-isometric Mappings. I, *Comm. Pure Appl.
    Math.*, 1968, 21, No 1, pp. 77–110.   (pp. 227, 228, 230, 232)

94. Kagan V.F., *Foundations of the Theory of Surfaces in Tensorial
    Exposition, Part I*, Moscow–Leningrad: GITTL, 1948.   (p. 94)

95. Kagan V.F., *Foundations of the Theory of Surfaces in Tensorial
    Exposition, Part II*, Moscow–Leningrad: GITTL, 1948. (p. 251)

96. Kaplan W., Close-to-convex Schlicht Functions, *Michigan
    Math. J.*, 1952, vol. 1, No 2, pp. 169–185.   (p. 52)

97. Kitagawa J., Periodicity of the Asymptotic Curves on Flat
    Tori in S^3, *J. Math. Soc. Japan*, 1988, vol. 40, No 3, pp. 457–476.
                                        (pp. 171, 197, 201, 205, 207)

98. Kitagawa Y., Rigidity of the Clifford Tori in S^3, *Math. Z.*, 1988,
    198, pp. 591–599.   (pp. 202, 203, 252)

99. Kitagawa Y., Embedded Flat Tori in the Unit 3-Sphere, *J.
    Math. Soc. Japan*, 1995, vol. 47, No 2, pp. 275–297.   (p. 179)

100. Kitagawa Y., An n-Dimensional Flat Torus in S^{2n-1} whose Ex-
    trinsic Diameter is Equal to π, *Kodai Math. J.*, 1997, 20, pp.
    156–160.   (p. 252)

101. Kitagawa Y., Isometric Deformations of Flat Tori in the 3-
    Sphere with Nonconstant Mean Curvature, *Tohoku Math. J.*,
    2000, vol. 52, pp. 283–298.   (p. 252)

102. Klingenberg W., *A Course in Differential Geometry*, Springer,
    1978.   (p. 112)

103. Kuiper N., On C^1-Isometric Imbeddings, I, II, *Nederl. Acad.
    Wetensch. Proc. Ser. A58, Indag. Math.*, 1955, 17, S. 545, 556,
    683–689.   (pp. 72, 195)

104. Lebesgue H., Integrale, longueur, aire, *Ann. di Matematica*, 1902, Ser. III, t. VII, pp. 231–359.   (pp. 6, 69, 70, 71, 72, 74)

105. Lebesgue H., Sur l'existence des plans tangents aux surfaces applicables sur le plan, *Fundamenta mathematicae*, 1935, vol. XXV, pp. 157–161.   (p. 71)

106. Lumiste Yu.G. and Chakmazyan A.V., Submanifolds with a Parallel Normal Vector Field, *Izv. Vuzov. Matematika*, 1974, No 5, pp. 148–157.   (p. 182)

107. Maschke H., Note on the Unilateral Surface of Moebius, *Trans. Amer. Math. Soc.*, 1900, vol. 1, p. 39.   (p. 172)

108. Massey W.S., Surfaces of Gaussian Curvature Zero in Euclidean Space, *Tohoku Math. J.*, 1962, vol. 14, No 1, pp. 73–79.   (pp. 85, 86)

109. Miklyukov V.M., Isothermic Coordinates on Surfaces with Singularities, *Mat. Sbornik*, 2004, vol. 195, No 1, pp. 69–88.   (p. 26)

110. Mishchenko A.S., Solovyev Yu.P. and Fomenko A.T., *Collection of Problems on Differential Geometry and Topology*, Moscow: Fizmatgiz, 2004.   (p. 188)

111. Mozharsky V.V., On Surfaces of Positive Curvature with the First-type Cuspidal Edge, *Ukr. Geom. Sbornik*, 1982, 25, pp. 110–116.   (p. 98)

112. Nash J., C^1-Isometric Imbeddings, *Ann. Math.*, 1954, 60, No 2, pp. 383–396.   (p. 72)

113. Nikolaev I.G., Parallel Translation and Smoothness of the Metric of Spaces with Bounded Curvature, *Dokl. Akad. Nauk SSSR*, 1980, vol. 250, No 5, pp. 1056–1058.   (p. 223)

114. Nikolaev I.G., On the Parallel Translation of Vectors in Spaces with Both-side Bounded Curvature, *Sib. Mat. J.*, 1983, vol. 24, No 1, pp. 130–145.   (p. 223)

115. Nikulin V.V. and Shafarevich I.R., *Geometry and Groups*, Moscow: Fizmatgiz, 1983. (p. 143)

116. O'Neill B., Isometric Immersions of Flat Riemannian Manifolds in Euclidean Space, *Michigan Math. J.*, 1962, 9, pp. 199–205. (pp. 179, 249)

117. Palais R., On the Differentiability of Isometries. *Proceed. of AMS*, 1957, vol. 8, No 4, pp. 805-807. (p. 20)

118. Pereslavskaya L.B., The Canonical Form of Monge–Ampère Parabolic Equations and Sufficient Conditions for the Existence of their Solutions Regular on the Whole Plane, *Usp. Mat. Nauk*, 1982, vol. 37, No 2, pp. 227–228. (p. 108)

119. Pereslavskaya L.B., Approximation of Solutions of Monge-Ampère Equations by Surfaces Amounting to Developable Ones, *Fund. Prikl. Mat.*, 2006, vol. 12, No 1, pp. 205–236. (p. 108)

120. Pereslavskaya L.B. and Rozendorn E.R., A Problem on the Velocity of Wind in the Subtropical Anticyclone, *Usp. Mat. Nauk*, 1995, vol. 50, No 4, p. 119. (p. 108)

121. Pereslavskaya L.B. and Rozendorn E.R., Monge-Ampère Parabolic Equations in Geometry and Meteorology, *Izv. Vuzov. Matematika*, 1996, No 2, pp. 41–43. (p. 108)

122. Pincall U., Hopf Tori in S^3, *Invent. Math.*, 1985, vol. 81, pp. 379–386. (pp. 179, 200, 201, 218)

123. Plastok R., Homeomorphisms between Banach Spaces, *Trans. Amer. Math. Soc.*, 1974, 200, pp. 169–183. (p. 225)

124. Pogorelov A.V., *Extrinsic Geometry of Convex Surfaces*, Moscow: Nauka, 1968 (English translation: *Transl. Math. Monogr.*, 35, Providence, R.I., 1973). (pp. 3, 73, 75, 80, 85, 90, 184, 191, 193)

125. Pogorelov A.V. *Bendings of Surfaces and Stability of Shells*, Kiev: Naukova Dumka, 1998. (p. 108)

126. Porciau B., Global Invertibility of Nonsmooth Mappings, *J. Math. Anal. Appl.*, 1988, 131, pp. 170–170.   (p. 228)

127. Poznyak E.G., Isometric Immersions of Two-dimensional Riemannian Metrics in Euclidean Spaces, *Usp. Mat. Nauk*, 1973, vol. 28, No 5, pp. 47–76.   (pp. 3, 179)

128. Poznyak E.G. and Shikin E.V., Surfaces of Negative Curvature, *Itogi Nauki i Tekhniki. Algebra. Topologiya. Geometriya*, vol. 12, Moscow: VINITI, 1974, pp. 171–207.   (pp. 3, 216)

129. Poznyak E.G. and Sokolov D.D., Isometric Immersions of Riemannian Spaces in Euclidean Spaces, *Itogi Nauki i Tekhniki. Algebra. Topologiya. Geometriya*, vol. 15, Moscow: VINITI, 1977, pp. 173–211.   (p. 3)

130. Rademacher H., Über partielle und total Differenzierbarkeit von Funktionen mehrer Variablen, und über die Transformation der Doppelintegrale, *Math. Ann.*, 1918, vol. 97, pp. 340–349. (p. 7)

131. Randrup T. and Røgen P., Sides of the Möbius Strip, *Arch. Math.*, 1996, vol. 66, pp. 511–521.   (pp. 150, 151, 155)

132. Randrup T. and Røgen P., How to Twist a Knot, *Arch. Math. (Basel)*, 1997, vol. 68, No.3, pp. 252–264.   (p. 154)

133. Røgen P., Embedding and Knotting of Flat Compact Surfaces in 3-Space, *Comment. Math. Helv.*, 2001, vol. 76, No 4, pp. 589–606.   (p. 144)

134. Reshetnyak Yu.G., Isothermic Coordinates in Manifolds of Bounded Curvature, Parts I and II, *Siber. Math. J.*, 1960, vol. 1, No 1, pp. 88–116 and vol. 1, No 2, pp. 248–276.
        (pp. 25, 26, 28, 37)

135. Reshetnyak Yu.G., *Stability Theorems in Geometry and Analysis*, Novosibirsk: Nauka, 1982 (English translation: Kluwer Acad. Publ., 1994).   (pp. 4, 19, 224, 232)

136. Reshetnyak Yu.G., Two-dimensional Manifolds of Bounded Curvature, *Itogi Nauki i Tekhniki. Sovr. Probl. Matematiki.*

268   I.Kh. SABITOV

Fund. Napravleniya, vol. 70, Geometriya-4, Moscow: VINITI, 1990, pp. 3–197 (English translation: *Encycl. Math. Sci.*, 70, Berlin: Springer, 1993, pp. 3–164).   (pp. 25, 26, 28, 70)

137. Reznikov A.G., Non-commutative Gauss Map, *Composito Math.*, 1992, 83, pp. 53–68.   (p. 201)

138. Rozendorn E.R., An Approximate Solution of the Equation of Wind–Pressure Relation for Anticyclones in Tropical and Subtropical Latitudes, *Dokl. Akad. Nauk SSSR*, 1980, vol. 253, No 3, pp. 584–587.   (p. 108)

139. Rozendorn E.R., The Equation of Wind–Pressure Relation in Models of Thin Atmosphere, *Vestn. Mosc. Univ. Math. Mech.*, 1981, No 4, pp. 11–14.   (p. 108)

140. Rozendorn E.R., A Class of Solutions of the Equation $z_{xx}z_{yy} - z_{xy}^2 + a\nabla z = 0$ and their Application to Some Problems of Meteorology, *Vestn. Mosk. Univ. Mat. Mekh.*, 1984, No 2, pp. 56–58.   (p. 108)

141. Rozendorn E.R., Surfaces of Negative Curvature, *Itogi Nauki i Tekhniki. Sovr. Probl. Matematiki. Fund. Napravleniya*, vol. 48, Geometriya-3, Moscow: VINITI, 1989, pp. 98–195 (English translation: *Encycl. Math. Sci.*, 48, Berlin: Springer, 1992, pp. 87–178).   (pp. 3, 108)

142. Sabitov I.Kh., On the Problem of Immersion of Two-dimensional Metrics into E^4, *Matem. Zametki*, 1977, vol. 21, No 2, pp. 137–140.   (p. 179)

143. Sabitov I.Kh., Isometric Immersion of Locally Euclidean Metrics in R^3, *Siber. Math. J.*, 1985, vol. 26, No 3, pp. 156–167.   (p. 120)

144. Sabitov I.Kh., On the Problem of Smoothness of Isometries, *Siber. Math. J.*, 1992, vol. 34, No 4, pp. 169–176.   (p. 4)

145. Sabitov I.Kh., Isometric Immersions and Embeddings of Locally Euclidean Metrics in R^2, *Izv. RAN. Matematika*, 1999, vol. 63, No 6, pp. 147–166.   (p. 28)

146. Sabitov I.Kh., Isometric Immersions and Embeddings of Flat Moebius Strip in Euclidean Spaces, *Izvestiya RAN. Matematika*, 2007, vol. 71, No. 5, pp. 197 – 224.   (pp. 151, 152, 170, 252)

147. Sabitov I.Kh., On Flexible Flat Tori in S^3. *Pure and Applied Differential Geometry PADGE 2007 (Eds. F. Dillen and I. Van de Woestyne)*, 2007, Aachen: Shaker Verlag, pp. 247 – 251.
(p. 202)

148. Sabitov I.Kh., Locally Euclidean Metrics with Given Geodesic Curvature of the Edge, *Trudy MIAN im. Steklova*, 2009, vol. 266 (in press).   (p. 66)

149. Sabitov I.Kh., On the Boundary of a Developable Surface with Small Regularity, *Siber. Math. J.*, 2009 (submitted).   (p. 111)

150. Sabitov I.Kh. and Shefel' S.Z., On the Relations between Orders of Smoothness of a Surface and its Metric, *Siber. Math. J.*, 1976, vol. 17, No 4, pp. 916–925.   (pp. 21, 22, 23, 147, 179)

151. Sacksteder R., The Rigidity of Hypersurfaces, *J. Mech.*, 1962, vol. 11, pp. 929–940.   (p. 252)

152. Sadovski M., Ein elementarer Beweis für die Existenz eines abwickelbaren Möbiusschen Bands und Züruckführung des geometrischen Problems auf ein Variationsproblem, *Sitzber. Preuss. Akad. Wiss.*, 1930, 22, S. 412–415.   (pp. 151, 155)

153. Sadovski M., Theorie der elastisch biegsamen undehnbaren Bänder mit Anwendung auf das Möbiussche Band, *Verh. 3 Kongr. Techn. Mechanik*, Stockholm, 1930, II, pp. 444–451.
(pp. 155, 156)

154. Sasaki S., On Complete Surfaces with Gaussian Curvature Zero in 3-Sphere, *Colloq. Math.*, 1972, vol. 26, pp. 165–174.
(pp. 179, 191)

155. Sasaki S., On Complete Flat Surfaces in Hyperbolic 3-Space, *Kodai Math. Sem. Reports*, 1973, vol. 25, No 4, pp. 449–457.
(p. 178)

270   I.Kh. SABITOV

156. Schwarz G., A Pretender to the Title "Canonical Moebius Strip", *Pacific J. Math.*, 1990, 143, pp. 195–200.
(pp. 150, 151, 155, 175)

157. Schwarz G., The Dark Side of the Moebius Strip, *Amer. Math. Monthly*, 1990, 97(10), pp. 890–897.   (pp. 150, 151, 156, 171)

158. Shefel' S.Z., Completely Regular Isometric Immersions in Euclidean Space, *Siber. Math. J.*, 1970, vol. 11, No 2, pp. 442–460.
(pp. 81, 82)

159. Shefel' S.Z., C^1-Smooth Isometric Immersions, *Siber. Math. J.*, 1974, vol. 15, No 6, pp. 1372–1393.   (pp. 76, 81, 93)

160. Shefel' S.Z., C^1-Smooth Surfaces of Bounded Exterior Positive Curvature, *Siber. Math. J.*, 1975, vol. 16, No 5, pp. 1122–1123.
(pp. 81, 82, 85)

161. Shtogrin M.I., Piecewise Smooth Embeddings of a Cube, *Usp. Math. Nauk*, 2004, vol. 59, No 5, pp. 167–168.   (p. 144)

162. Shtogrin M.I., Special Isometric Transformations of Surfaces of Plato's Bodies, *Usp. Math. Nauk*, 2005, vol. 60, No 4, pp. 221–222.   (pp. 144, 252)

163. Shtogrin M.I., Piecewise Smooth Developable Surfaces, *Trudy MIAN im. Steklova*, 2008, vol. 263, pp. 227–250.   (p. 106)

164. Shtogrin M.I., Isometric Immersions of a Cone and a Cylinder, *Izv. RAN. Matematika*, 2009, vol. 73, No. 1, pp. 187–224.
(p. 252)

165. Shtogrin M.I., Bendings of Developable Surfaces with Preservation of the Edge and Generatrices, *Trudy MIAN im. Steklova*, 2008, vol. 266 (in press).   (p. 252)

166. Spivak M., *A Comprehensive Intruduction to Differential Geometry*, vol. IV, Boston: Publish or Perish Inc., 1975.
(pp. 178, 179, 180, 182, 184, 191, 200)

167. Spivak M., Some Left-over Problems from Classical Differential Geometry, *Proc. Symp. Pure Math.*, 1975, vol. 27, pp. 245–252.
(pp. 179, 184)

168. Starostin E.L. and G.H.M. van der Heijden, The Shape of a Möbius Strip, *Nature Materials*, 2007, 6(8), pp. 563–567. DOI:10.1038/nmat1929.   (p. 156)

169. Starostin E.L. and G.H.M. van der Heijden, The Equilibrium Shape of an Elastic Developable Möbius Strip, *PAMM, Proc. Appl. Math. Mech.*, 2007, 7(1) pp. 2020115–6. DOI: 10.1002/pamm.200700858.   (p. 156)

170. Stiel E., Isometric Immersions of Manifolds of Nonnegative Constant Sectional Curvature, *Pacif. J. Math.*, 1965, vol. 15, No 4, pp. 1415–1419.   (pp. 179, 249)

171. Stiel E., Immersions into Manifolds of Negative Curvature, *Proc. Amer. Math. Soc.*, 1967, vol. 18, pp. 713–715.   (p. 249)

172. Stoker J.J., Developable Surfaces in the Large, *Commun. Pure Appl. Math.*, 1961, vol. XIV, pp. 627–635.   (pp. 73, 85, 86, 87)

173. Stoker J.J., Uniquness Theorems for Surfaces of Cones or Cylinders Closed off at One End by Convex Bases, *Commun. Pure Appl. Math.*, 1970, vol. XXIII, pp. 425–446.   (pp. 73, 252)

174. Titus C.J., A Theory of Normal Curves and Some Applications, *Pacif. J. Math.*, 1960, 10, No 3, pp. 1083–1096.   (p. 52)

175. Tompkins C., Isometric Embedding of a Flat Manifold in Euclidean Space, *Duke Math. J.*, 1939, 5, pp. 58–61.   (p. 248)

176. Tompkins C. A Flat Klein Bottle Isometrically Embedded in Euclidean 4-Space, *Bull. Amer. Math. Soc.*, 1941, 47, p. 208.
(pp. 179, 214)

177 Toponogov V.A., Riemannian Spaces which Contain Straight Lines, *Dokl. Akad. Nauk SSSR*, 1959, vol. 127, No 5, pp. 977–979.   (p. 225)

178. Toponogov V.A., The Metric Structure of Riemannian Spaces of Nonegative Curvature Containing Straight Lines, *Siber. Math. J.*, 1964, vol. 5, No 6, pp. 1358–1369.   (pp. 91, 225)

179. Ushakov V.G., *Riemannian Manifolds and Surfaces of Constant Nullity*, PhD Thesis, St.-Petersbourg University, 1993.   (pp. 235, 236, 239, 240, 241, 247)

180. Ushakov V., Parametrization of Developable Surfaces by Asymptotic Lines, *Bull. Austral. Math. Soc.*, 1996, 54, pp. 411–421.   (pp. 112, 113)

181. Ushakov V., Smoothness of the Solution of Trivial Monge–Ampère Equation, *Bull. Austral. Math. Soc.*, 1997, 56, pp. 439–445.   (pp. 94, 95, 99)

182. Ushakov V., Developable Surfaces in Euclidean Space, *J. Austral. Math. Soc.*, 1999, 66, pp. 388–402.   (pp. 69, 236, 239)

183. Ushakov V., The Explicit General Solution of Trivial Monge–Ampère Equation, *Comment. Math. Helvetici*, 2000, 75, pp. 125–133.   (pp. 94, 95, 100, 101, 102)

184. Vekua I.N., *Generalized Analytic Functions*, Moscow: Fizmatgiz, 1959 (2nd edn, 1988).   (pp. 9, 10, 11, 26, 32, 41, 43, 44, 47, 49, 57, 120)

185. Vinogradsky A.S., Causes for Nonextensibility of Surfaces of Negative Curvature Bounded away from Zero over the Smooth Boundary, *Vestn. Mosk. Univ. Mat. Mekh.*, 1970, No 5, pp. 83–86.   (p. 97)

186. Volkov Yu.A. and Vladimirova C.M., Isometric Immersions of Euclidean Plane into Lobachevsky Space, *Mat. Zametki*, 1971, vol. 10, No 3, pp. 327–332.   (p. 178)

187. Weiner J.L., Flat Tori in S^3 and their Gauss Map, *Proc. London Math. Soc.*, 1991, vol. 62, No 3, pp. 54–76.   (pp. 179, 201, 207)

188. Weiner J.L., *Rigidity of Clifford Tori. Geometry and Topology of Submanifolds, VII (Leuven, 1994; Brussels, 1994)*, World Sci. Publishing, River Edge, NJ, 1995, pp. 274–277.   (pp. 203, 252)

189. Weiner J.L., Isometric Immersions of E^2 into E^4, in: *Geometry of Submanifolds and Related Topics*, Kyoto: Surikaisekikenkyusho Kokyuoroku, 2001, pp. 136–143.   (pp. 179, 183)

190. Wintner A., On Riemann Metrics of Constant Curvature, *Amer. Math. J.*, 1951, 73, No 2, pp. 569–575.
        (pp. 145, 219, 222)

191. Whitt L., Codimension Two Isometric Immersions between Euclidean Spaces, *Pacific J. Math.*, 1985, 115, pp. 481–487.
        (p. 179)

192. Wolf J.A., *Spaces of Constant Curvature*, University of California, Berkeley, 1972.   (pp. 143, 217, 224)

193. Wunderlich W., Über ein abwickelbares Möbiusband, *Monats. für Mathematik*, 1962, Band 66, Heft 3, S. 276–289.
        (pp. 150, 155, 176)

194. Yanenko N.N., The Geometric Structure of Small-type Surfaces, *Dokl. Akad. Nauk SSSR*, 1949, vol. 64, No 5, pp. 641–644.
        (pp. 179, 180)

195. Yanenko N.N., Some Questions of the Theory of Embeddings of Riemannian Metrics into Euclidean Spaces, *Usp. Mat. Nauk*, 1953, vol. 8, No 1, pp. 21–100.   (pp. 179, 180)

196. Yao S.T., Submanifolds of Constant Mean Curvature, II, *Amer. Math. J.*, 1975, vol. 97, No. 1, pp. 76–100.   (p. 201)

197. Zalgaller V.A., On the Singularities of C^1-Smooth Surfaces, *Vestn. Leningr. Univ.*, 1962, 7:2, pp. 71–77.
        (pp. 19, 72, 145, 147)

198. Zalgaller V.A., *Theory of Envelopes*, Moscow: Fizmatgiz, 1975.
        (p. 103)

199. Zhou Z.M. and Chen J. X., Quasiconformal Mappings with Nondecreasable Beltrami Coefficients, *Acta Math. Sin.*, 2003, vol. 46, No 2, pp. 379–384.   (p. 26)

INDEX

REVIEWS IN MATHEMATICS AND MATHEMATICAL PHYSICS

Editor:
A.T. Fomenko
Moscow State University
Russia

Aims and Scope
Reviews in Mathematics and Mathematical Physics publishes review papers covering significant developments in mathematics and mathematical physics from Russia and the former Soviet Union. Topics include soliton theory, the theory of quantum topological models and their applications for algebraic and differential geometry and topology.

World Wide Web Address
Additional information is also available through the Publisher's web home page site at http://www.cambridgescientificpublishers.com.

Ordering Information
Each volume consists of an irregular number of parts depending upon extent. Issues are available individually as well as by subscription. 2008/2009 Volume 13/14.

Orders may be placed with your usual supplier or at the address shown below. Journal subscriptions are sold on a per volume basis only. Claims for nonreceipt of issues will be honored if made within three months of publication of the issue. All issues are dispatched by airmail throughout the world.

Subscription Rates
Base list subscription price per volume: EUR 80.00.* This price is available only to individuals whose library subscribes to the journal OR who warrant that the journal is for their own use and provide a home address for mailing. Orders must be sent directly to the Publisher and payment must be made by check or credit card.

Separate rates apply to academic and corporate/government institutions.

*EUR (Euro). The Euro is the worldwide base list currency rate. All other currency payments should be made using the current conversion rate set by Publisher. Subscribers should contact their agents of the Publisher. All prices are subject to change without notice.

Reviews in Mathematics and Mathematical Physics

Notes for Contributors

Manuscripts
Manuscripts submitted as hard copy should be typed with double spacing and wide margins (3 cm) on good quality paper, and submitted in triplicate to A.T. Fomenko, Editor, Reviews in Mathematics and Mathematical Physics (Chair of Differential Geometry and Applications, Department of Mathematics and Mechanics, Moscow State University, 119899, Russia. Email: fomenko@orc.ru) or sent to Janie Wardle, Cambridge Scientific Publishers, P.O. Box 806, Cottenham, Cambridge, CB24 8RT, UK. Email: janie.wardle@cambridgescientificpublishers.com.

Submission of a paper to *Reviews in Mathematics and Mathematical Physics* will be taken to imply that it represents original work not previously published, that it is not being considered for publication elsewhere, and that if accepted for publication it will not be published elsewhere in the same language without the consent of the Editor and Publisher.

Language: The language of publication is English, and manuscripts should be submitted in English if possible, although Russian manuscripts can be accepted.

Abstract: Each paper requires an abstract of 100–150 words summarizing the significant coverage and findings. It is a condition of acceptance by the Editor of a typescript for publication that the Publishers acquire automatically the copyright in the typescript throughout the world. Key words, mathematics subject classification and running head should also be included.

Figures
All figures should be numbered with consecutive Arabic numbers, have descriptive captions and be mentioned in the text. Keep figures separate from the text, but indicate an approximate position for each in the margin.

Preparation: Figures submitted must be of a high enough standard for direct reproduction. Line drawings should be prepared in black (India) ink on white paper or on tracing cloth, with all the lettering and symbols included. Alternatively, good sharp photoprints ("glossies") are acceptable. Clearly label each figure with author's name and figure number, indicate "top" where this is not obvious. Redrawing or retouching of unusable figures will be charged to the authors.

Size: Figures should be planned so that they reduce to 11 cm column width. The preferred width of line drawings is 15 to 22 cm with capital lettering 4 mm high, for reduction by one-half.

Color Plates: Whenever the use of color is an integral part of the research, or where the work is generated in color, the Journal will publish the color illustrations without charge to the author. Reprints in color will carry a surcharge. Please write to the Publisher for details.

Equations and Formulae
Mathematical notation should be clearly marked.

Mathematical equations should preferably be typewritten, with subscripts and superscripts clearly shown. It is helpful to identify unusual or ambiguous symbols in the margin when they first occur. To simplify typesetting, please use: 1) the "exp" form of complex exponential functions; 2) fractional exponents instead of root signs; and 3) the solidus (/) to simplify fractions – e.g. $\exp x^{1/2}$. Please underline all mathematical symbols to be set in italic and put a wavy line under bold symbols. Other letters not marked will be set in roman type.

Tables

Number tables consecutively with Arabic numerals and give a clear descriptive caption at the top. Avoid the use of vertical rules in the tables. Indicate in the margin where the printer should place the tables.

References and Notes

References and notes are indicated in the text by consecutive Arabic numbers (with parentheses). The full list should be collected and typed at the end of the paper possibly in alphabetical order. Use of unpublished results in the proofs should be avoided.

Listed references should be complete in all details. Authors' initials should precede their names; journal title abbreviations should follow style of mathematical Reviews Index.

The authors should be aware that references will not be admitted on unpublished papers and books.

Examples:

[1] G.N. Watson, A treatise on the theory of Bessel functions. 2nd ed. Cambridge Univ. Press, London and New York, 1945.

[2] H. Mellin, Abriss einer enheitlichen Theorie der Gamma und der hypergeometrischen Functionen, Math. Ann. 68 (1910), 305-337.

Proofs

Russian contributors will receive page proofs (including figures) for correction via our internal courier network to Moscow. These must be returned to our Moscow office (Victor Selivanov, Room 123, Lebedev Physical Institute, 53 Leninsky Prospect, Moscow 117924, Russia. Email: selivanov@sci.lebedev.ru) within 48 hours of receipt. All other contributors will receive page proofs (including figures) by airmail for correction, which must be returned to the printer within 48 hours of receipt. Please ensure that a full postal address is given on the first page of the typescript, so that proofs are not delayed in the post. Author's alterations in excess of 10% of the original composition cost will be charged to authors.

Reprints

Additional reprints may be ordered by completing the appropriate form sent with proofs.

Page Charges

There are no page charges to individuals or institutions.

Lightning Source UK Ltd.
Milton Keynes UK
04 January 2010

148158UK00001B/95/P